# GEOGRAPHY OF INDIA

# GEOGRAPHY OF INDIA

## COMPRISING A DESCRIPTIVE OUTLINE OF ALL INDIA,

## AND A DETAILED

## GEOGRAPHICAL, COMMERCIAL, SOCIAL, AND POLITICAL ACCOUNT OF EACH OF ITS PROVINCES

GEORGE DUNCAN

GYAN

*Published by*

Gyan Publishing House
5, Ansari Road
Daryaganj, New Delhi-110002
Phone: 011-47034999, 9811692060
E-mail: books@gyanbooks.com

*Distribution Network*
gyanbooks.com
India, USA, Canada, UK, Australia



ISBN: 978-93-7169-456-8 (PB)

First Published, 1880

2nd Impression 2025

*Printed at:* Gyan Press, Delhi.



**GEOGRAPHY OF INDIA**
Author: GEORGE DUNCAN

Buddhist Monastery, Himálayas.

# A GEOGRAPHY OF INDIA,

COMPRISING A DESCRIPTIVE OUTLINE OF ALL INDIA,

AND A DETAILED

GEOGRAPHICAL, COMMERCIAL, SOCIAL, AND POLITICAL ACCOUNT OF EACH OF ITS PROVINCES.

*WITH HISTORICAL NOTES,*

BY

GEORGE DUNCAN.

Eleventh Edition.

(*Revised and corrected to date from the latest Official Information.*)

LONDON:
TRÜBNER & CO., LUDGATE HILL.
1881.

# NOTE.

In pronouncing Indian names, the reader will bear in mind that *a* is sounded as *a* in *cedar*, or as *u* in *gun*; *á* as in *father*; *e* as *a* in *made*; *i* as in *bit*; *í* as *i* in *machine*; *o* as in *hole*; *u* as in *full*; *ú* as *u* in *rude*; *ai* as it is in *aisle*; *au* as *ou* in *house*; *ch* as in *church*; and *th* as *t* in *at*.

In respect to the attempt to give the true spelling of the proper names used in this book, it may be stated that where a name occurs in two forms, thus, Attock (Atak), Benares (Banáras), the one added in a parenthesis is that which most nearly approaches the local vernacular, and should be recognised in preference to the first or more common anglicised form. When no second mode of spelling a name occurs, the one given is either (1) the equivalent of the vernacular, *e.g.*, Panjáb, and the mode now generally adopted, or (2) is a form so fixed by usage, *e.g.*, Calcutta, that it is not likely ever to be disturbed.

G. D.

# CONTENTS.

Ancient Hindu Sculpture. Brahma, "the Lotus-Born."

A

# GEOGRAPHY OF INDIA.

1. *Position.*—India lies between N. lat. 8° 4′—36°, E. long. 66° 44′—99° 30′.

2. *Boundaries.*—The natural boundaries of India are remarkably well defined. The whole of the northern frontier is formed by the gigantic range of the Himálayas; and the southern, or rather the south-eastern and south-western limits, are fixed by the Indian Ocean. The Sulaimán and Hála ranges of mountains separate it from Afghánistán and Baluchistán on the west, and long lines of watershed bound it on the east.

3. *Dimensions.*—Its greatest length, measured from Cape Comorin in the south to the extremity of the Panjáb in the north, is estimated at 1800 miles, and its breadth, from Karáchi in the west to the extremity of Assam (Asam) in the east, at about the same distance. Its superficial area is 1,400,000 square miles. The coast-line exceeds 4000 miles in length.

4. *Physical Features.*—India presents all the physical features of a vast continent, and in wonderful variety, from the swamps and forests of the Tarái to the parched-

up desert of the Thar, from the rich soil of Bengal to the salt wastes of Kach, from the dull level of the plains to the lovely scenery of the Western Gháts, and from the low-lying Carnatic (Karnátik) to the snow-clad peaks of Himálaya. Roughly speaking, however, India may be divided into *five* distinct regions:—(1) the vast **Mountain Mass of the Himálayas** on the north; (2) the **Indo-Gangetic Plain,** or plain of the Indus and the Ganges, reaching from the Arabian Sea on the west to the Bay of Bengal on the east; (3) the **Plateau of Málwa,** extending in a triangular form near Delhi on the north to the Vindhya Mountains on the south; (4) the **Tableland of the Dakhan,** bounded on the east and west by the Eastern and Western Gháts; and (5) the **Coast Plains** below the Gháts.

5. MOUNTAINS.—The **Himálaya** Mountains extend over a length of 1500 miles, and are on the average 150 in breadth. They stretch from the defile above Kashmír on the north-west, through which the Indus penetrates into the plains of the Panjáb, to the southern bend of the Sanpu on the east. The mean height is 20,000, and that of the snow-line 18,500 feet.

[Many of the peaks rise to great elevations. Mount Everest, the highest mountain in the world, whose summit overtops that of the principal peak of the Andes by half a mile in perpendicular height, is 29,000 feet above the sea; Kanchinjinga 28,100, and Daulagiri 28,000. The lower parts, or PASSES, as they are called, from their affording to travellers the means of passage from one side to the other, are rarely under 17,000 feet above the sea. Viewed from the plains, the Himálayas present long ridges of hills intersected by valleys, rising one above another until they are lost in the clouds.]

6. The **Sewálik** Mountains run parallel to the Himálayas between the Ganges and the Satlej.

[The highest point is 3500 feet above the sea, and where widest they are ten miles broad. Nearing the Satlej they sink into mere sandhills. The range is remarkable for the enormous quantities of fossil remains of gigantic animals which are found in it.]

7. The **Arávali** Mountains are a series of ridges of hills between the basins of the Indus and the Ganges.

[They extend from S.W. to N.E., more than 300 miles, with a breadth of from six to sixty miles, and a general elevation of 3600 feet.]

8. The **Vindhya** * range is separated from the Arávalis by the valley of the Máhi.

[It extends east and west for about 350 miles, and terminates in the hilly region of Central India. The Vindhyas are not quite so high as the Arávalis. On the northern side they form part of the tableland of Málwa, 2000 feet above the level of the sea.]

9. The **Sátpura** Mountains run parallel to the Vindhya hills, the two ranges being separated only by the valley of the Narbada (Narmadá). South of these again, and of the valley of the Taptí, rise the **Chandur** and **Gawalgarh** ranges, forming the northern border of the plateau of the Dakhan. The **Western Gháts** extend from near the mouth of the Taptí to Cape Comorin.

[These hills rise abruptly at about thirty or forty miles from the coast, forming on the eastern side an undulating tableland 2000 feet above the level of the sea. The northern part of the chain, known as the Sáhyádri Hills, does not exceed 3000 feet in height; at Mahábaléshwar the chain rises to 4500 feet, and at the Nílgiris to nearly 9000 feet.]

10. The **Eastern Gháts** are a much less elevated

---

* *Vindhya,* "the barrier," so called from its being near the Tropic of Cancer,—the barrier, as it were, to the sun's progress in his northern course.

range than the Western Gháts, are much more broken, and are situated at a much greater distance from the sea. The chain is crossed by all the rivers which flow from the interior tableland. The most important portion is called the **Shervarái** Hills, 4500 feet high. Linking the Eastern and Western Gháts together at their southern extremities are the **Palnai** Hills, 7000 feet high.

11. On the east are the **Patkoi** Hills; the **Yoma** Mountains, forming the eastern boundary of Arakan; and the **Siamese** Mountains, separating Tenasserim from Siam.

12. The **Sulaimán** range on the west of the Indus runs from north to south for 350 miles. The greatest elevation is 11,000 feet above the sea.

[The sides are clothed with dense and lofty forests, reaching nearly to the top; and the valleys are overgrown with trees, shrubs, and flowers indigenous to the soil.]

13. *Rivers.*—The **Indus** rises in Mount Kailás on the north side of the Himálayas, 22,000 feet above the sea. It flows generally in a north-western direction, until at the point where it leaves the mountains it takes a bend to the south and keeps that course to the Arabian Sea, into which it flows, after a course of 1800 miles, by many channels and branches. The towns upon or near its banks are **Haidarábád, Lárkhána, Rohri, Dera Gházi Khán, Derah Ismáil Khán,** and **Attock** (Atak).

[In the first 360 miles of its length the Indus falls at the rate of 22 feet per mile. From Attock (Atak), where it begins to be navigable, to the sea, a distance of 940 miles, it falls at the rate of one foot per mile. The breadth, depth, and velocity are very inconstant throughout; in some parts the river flows in narrow channels confined by lofty wall-like rocks, in others inundates the surrounding country for many miles; in some places it is 200 feet deep, at others not half as many inches; sometimes it dashes along with irresistible fury, at others slowly meanders onward to the sea. Numerous islands are formed in its course, and the navigation is at all times difficult. It is crossed at Attock (Atak) by a bridge of boats, and at several other places by steam-ferries. Steamers constructed for the purpose navigate the river at all seasons. Throughout its course it receives many large streams on both sides; on the right bank the *Shayok*, and the *Kábul*, navigable to within fifty miles of Kábul; and on the left the *Panjnad*, and many others. This last affluent flows into the Indus near Mithankot, and is composed of the five celebrated rivers of the Panjáb,—the *Jhílam*, the *Chenáb*, the *Rávi*, the *Biás*, and the *Satlej*. The first four have their sources on the south side of the Himálayas; the Satlej rises on the north. They are all very large navigable rivers, but much obstructed by sandbanks.]

14. The **Ganges** is formed by the junction of the **Bhágirathi** and the **Alaknanda**, which unite at Deoprayág in Garhwal, 1500 feet above the sea. Thirty miles below the junction, the river passes Hardwár, 1000 feet above the sea, and entering the great valley of Hindustán, flows in a S.E. direction into the Bay of Bengal. It is joined on the right by the **Jamna** (which receives the **Chambal**) and the **Són**, and on the left by the **Rámganga**, **Gumtí** (Gomatí), **Gogra**, **Gandak** (Gandakí), and **Kosi**.* The chief towns on or near its banks are **Farrukhábád**, **Cawnpore** (Kánhpur), **Allahábád**, **Benares** (Banáras), **Patna**, **Murshidábád**, and **Calcutta**.

[The delta commences about 200 miles from the sea, and the southern portion, known as the Súndarbans, is a dreary, unhealthy

* In Sanskrit *Kusiki* or *Kausiki*, so called (according to the Puránas) from Kusiki, daughter of Raja Kusika (= squint-eyed), and wife of a pious but irascible Bráhman sage, who in a fit of anger changed her into this river.

region, broken up by creeks and rivers into innumerable swampy islands covered with jungle and infested with wild beasts. Owing to the shifting of the mudbanks, the navigation of the delta is very dangerous, the *Hugli* being the only branch which can be used with safety by ships of large burthen. The Ganges is about three miles wide in its broadest part, and in the dry season about thirty feet deep where deepest; but by the end of July, being swollen with the rains, it spreads over all the flat country of Bengal. The valley of the Ganges is one of the richest on the globe, the soil consisting entirely of vegetable mould. Wheat is produced in the upper part of the valley, and rice, cotton, indigo, opium, and fruit of all kinds are grown in the lower. The length of the Ganges is 1600 miles; it is navigable for boats as far as Hardwár, 1400 miles from the sea. Along the whole of this distance the traffic is enormous, but below Allahábád the river is crowded with craft of all descriptions. The navigation of the river is assisted by a canal from Cawnpore (Kánhpur) to Hardwár, which, with its branches, serves also to irrigate the whole of the country between the Ganges and the Jamna.

The Ganges is subject to the phenomenon called the *bore.* On the rise of the tide the river near the sea sometimes rises suddenly to the height of ten feet, and rushes onward with great noise at the rate of seventeen or eighteen miles an hour, subsiding as it goes. At Calcutta the height of the bore is often five feet.

The Hindus hold the water of the Ganges sacred from Gangautri (Ganga avatári), about fifteen miles from its source, to Ságar, an island at the mouth of the Hugli. There are, however, particular portions held more sacred than the rest, to which pilgrims resort from all parts of India to perform their ablutions and to carry off the water to be used in future ceremonies.]

15. The **Brahmaputra** (= Son of Brahma) rises on the north slope of the Himálayas, nearly opposite the sources of the Ganges and not far from those of the Indus. During the first part of its course it is called the **Sanpu**, and flows in an easterly direction parallel to the Himálayas, when it suddenly turns southward forcing a passage through the mountains, and enters Assam (Asam) under the name of the **Dihong**. It is afterwards called

**Brahmaputra**, and keeps that name till within a comparatively short distance of the sea, when it is changed to **Megna**. The length of the river is 1400 miles, and the volume of water discharged by it nearly twice as great as that discharged by the Ganges. The principal tributaries are the **Dikhu**, **Surmá** or **Barák**, **Gumtí**, **Nilkomar**, and **Tísta**.

[During the floods from June to September the plains of Assam (Asam) are one sheet of water, and the Brahmaputra unites with the Ganges in inundating the low grounds of Bengal. In its course through Assam (Asam) and Bengal it forms numerous islands, and as it approaches the sea its stream is four or five miles wide. Its waters are very muddy, and its banks covered with jungle or marshes. Like the Hugli, it is subject to the sudden rise of the tide called the bore.]

16. The **Mahánadi** (= Great River) rises in Nauagada, one of the native states on the south-west frontier of Bengal, and after an irregular course of 520 miles discharges itself by many mouths into the Bay of Bengal.

[At Sambalpur, 260 miles above its mouth, it is nearly a mile wide, and near Cuttack (Katak), where the delta begins, quite two miles in breadth.]

17. The **Godávari** rises near Násik in Ahmadnagar, on the eastern slope of the Western Gháts, and flowing in a general easterly direction through the Nizám's dominions as far as Sironcha, and south-easterly between that territory and the Central Provinces, separates into two wide and many smaller branches near Rajahmundry (Rájámáhendripuram), and after a course of 900 miles falls into the Bay of Bengal. The tributaries are on the right the **Manjira**, and on the left the **Purna**, the **Pranhíta** (formed by the **Páinganga** from the Nizám's

territories, and the **Wáinganga** from the Central Provinces), and the **Indráwati**.

[Rocky barriers interrupt the navigation of the Godávari, but it is practicable for boats at all times of the year. Timber rafts from the forests of Nágpur are floated down to the sea. The country lying between the Jamna and the Godávari comprised the ancient forest of *Dandaka*, which in the "Rámáyana" is described as inhabited by savages, whom the poet calls "Rákshasas," and infested with wild beasts.]

18. The River **Krishna** (= Black River) rises on the eastern slope of the Western Gháts at Máhábaléshwar, forty miles from the western coast, and flowing across the peninsula in an easterly direction, enters the Bay of Bengal by several branches after a course of 800 miles. On the right bank it receives the **Málprabha** and the **Tungabhadra**, and on the left the **Bhíma**.

[The banks of the Krishna throughout its course are higher than the surrounding country, and except at the delta the river is useless for purposes of irrigation. It is also quite unnavigable, owing to the rapidity of its current and the rockiness of its channel.]

19. The **Pennér** (the Sanskrit *Pinákiní*) rises in a tank in the fort of Chandradrug in Mysore (Maisúr), and after a course of 355 miles falls into the Bay of Bengal. The **Pálár** rises in Mysore (Maisúr), and after a course (generally south-east) of 220 miles falls into the Bay of Bengal a little south of Sadras (Sathurangapattanam). The **Pennár** rises north of the Nandidrug Hills in Mysore (Maisúr), and after a course of 244 miles enters the Bay of Bengal a mile north of Fort St. David.

NOTE.—The Pennér, the Pálár, and the Pennár are all dry in the hot season.

20. The **Káveri** * rises in Coorg (Kodagu) and flows south-easterly through Mysore (Maisúr), then southward, forming the boundary between Salem (Sélam) and Coimbatore (Kóyambuttúr), then east through Trichinopoly (Tiruchirápalli) and Tanjore (Tanjávúr) to the sea. The entire length of the Káveri is 472 miles.

[At Trichinopoly (Tiruchirápalli) the Káveri divides into several streams, embracing a delta seventy miles in length from the apex to the sea. The most important branch is the Coleroon (Kolladam), many of the other branches losing themselves in numberless cuttings and channels before reaching the coast. Numerous tributaries flow into the Káveri, but they are all, together with the Káveri itself, in the dry season from March to May, little more than watercourses. During the heavy rains, however, brought by the S.W. monsoon in June and July, they are filled to overflowing, so that in August all the low grounds are inundated. The soil brought down by the river renders its delta one of the most fertile parts of India. The Káveri is not navigable except down-stream by rafts formed of waterpots and planks, or skins stretched over bamboos; cotton, sandalwood, grain, and tamarinds are brought down in this manner.]

21. The **Taptí** rises near Baitul, and flows west with a very winding course of about 500 miles into the Gulf of Kambay (Kambáyat) near Súrat. The **Narbada** (Narmadá, Sanskrit *narma* = delight, *da* = giving) rises in the elevated plateau of Amarakantak in the district of Rámgarh near the source of the Són. It flows at first along the tableland; has a great fall near Mandla; and then turning nearly due west between the Vindhya and the Sátpura Mountains, falls into the Gulf of

* According to local tradition, on one occasion, when the rishi Agastia Muni was absorbed in the contemplation of Bráhma, a crow perched on his waterpot and overturned it. The water streamed from the vessel, and never ceasing to flow from it, gives rise to the Káveri.

Kambay (Kambáyat). It has a course of 750 miles, with fewer windings than most Indian rivers; but it is obstructed by rocks, shallows, and cataracts.

22. *Lakes.*—Unlike the mountains and rivers of India, which are on a scale of the greatest magnitude, the lakes are few and of small dimensions. They are **Chilká** Lake, on the coast of Orissa; **Kolair** (Kollér) Lake, between the Godávari and the Krishna; **Pulicat** Lake, a little north of Madras; and the salt lakes of **Sámbar, Didwána,** and **Sír** in Rájputána. The **Ran** of **Kach** is a lake only in the rainy season.

23. *Gulfs, Bays, and Straits.*—The **Gulf of Kach,** the **Gulf of Kambay,** the **Gulf of Mannár, Pámban Passage, Palk's Straits, False Bay, Bálasor Roads**

24. *Capes.*—**Jagat Point,** the western point, and **Diu Head,** the southern point of the peninsula of Kathiawár, in Gujarát; **Cape Comorin** (Kanyá Kumárí), the most southern point in India; **Point Kalimír,** the **Dolphin's Nose,** the **Sandheads,** and **Cape Negrais.**

25. *Climate.*—As India extends through 28° of latitude and is greatly varied in respect of elevation, it necessarily presents much variety of climate. The heat in the plains from May to August is often intense, while the temperature of some of the more elevated districts during the same period is delightfully cool. The quantity of rain which falls in any particular locality depends greatly on its position with respect to the east and the west coasts, that is, to the amount of its exposure to the influence of either or both of the monsoons.

[Over the peninsula of India the winds make an annual tour with the regularity almost of clockwork. On the Coromandel coast they blow from the north-east in November, and directly from the east about the end of February, then from the south-east, gradually going south, until by the middle of April they blow directly from the south; veering round towards the west, about the end of May the winds blow from the south-west and west, bringing with them masses of rain-charged clouds, which burst on the coast about the beginning of June. The rainfall is heaviest at the western end of the Nílgiris, on the Kundas, which are the highest parts of the line of Gháts, and in Coorg (Kodagu). At Calicut (Kallikód) the average fall is 120 inches, and at Mangalúr 180. Thence in travelling northwards the rain gradually decreases in amount, and passing into Gujarát still diminishing, is lost altogether in Sindh, where there is no monsoon. The clouds which pass to the south of Cape Comorin traverse the Bay of Bengal, until coming into contact with the high mountain ranges in the eastern peninsula, they are diverted from a north-easterly to a north-westerly course, and so the rain descends in Bengal, gradually diminishing in quantity as it passes along the southern face of the Himálayas, till in Afghánistán it falls only in occasional showers. Thus the periodical rains begin at Calicut (Kallikód) and Calcutta nearly at the same time. In August the S.W. wind begins to shift round to the north-east, and the rain bursts on the east coast about the end of October, but in much less quantity than that which is carried to the western coast by the S.W. monsoon.]

[*The Monsoons.*—The periodical winds which blow over the Arabian, Indian, and China Seas, from about 10° south of the equator to 30° north of it, are called "monsoons." If air be heated, it will expand and mount upwards, and cold air from without will rush in to supply its place. This process carried out on an enormous scale is the main cause of the monsoons. From April to October, when the sun's rays fall with their greatest energy on that part of the earth lying between the equator and the Tropic of Cancer, the extensive countries of Arabia, India, and China, and their adjacent seas, becoming heated, the air in contact with them becomes heated also, so that it rises and gives place to fresh supplies drawn from the equator. In a similar way, from October to May, when the sun's rays fall with greatest effect on the region south of the Line, cool air rushes southwards to fill the space left by that which has become heated. Thus the monsoons would not blow from the north-east and south-west, as they do now, but directly from north and south, if they were caused solely by the action of the sun. The earth is a spherical body rotating from west to east; the air surrounding it travels along with it, consequently the atmosphere at the equator travels with a much greater velocity than that of the countries and seas farther removed from it, and therefore when air is drawn northward from the equator, having also a strong tendency to go eastward, the direction it takes is north-east. Wind blowing *towards* the north-east is called a south-west wind, and hence the "south-west monsoon." On the other hand, when colder air is drawn from Arabia, India, and China to the regions of the equator, it travels from slower-moving to quicker-moving parts of the earth, and consequently lags more and more to westward as it approaches the Line; hence the "north-east monsoon."]

26. *People.*—The population of India is estimated at 240 **millions**, and is composed of many races, differing considerably in appearance, dress, language, customs, and character in different localities; as, for example, the Bengalis, the Rájputs, and the Maráthas. Aboriginal tribes, distinct from the people of the plains and from each other, occupy the hills and forests.

27. *Religion.*—By far the greater number of the people of India profess **Bráhmanism**; other prevalent religions are **Muhammadanism** and **Buddhism**, the **Jain** and the **Sikh** religions, and **Christianity**.

28. *Language.*—There are at least thirty distinct languages in India, and of these there are many diversities of dialect. **Bengalí** is spoken in Bengal; **Hindustání** is the chief vernacular of the Muhammadans throughout the country, and **Hindí** that of the Hindu population of the North-West Provinces; **Assamese**, **Nepálese**, **Kashmírí**, **Panjábí**, **Gujarátí**, and **Kachí**, are spoken, as their names intimate, in Assam (Asam),

Nepál, Kashmír, the Panjáb, Gujarát, and Kach respectively; all these are derived from the Sanskrit. **Sindhí** is spoken in Sindh, and **Máráthí** by the Máráthas. In Southern India **Tamil** is the vernacular language of the people inhabiting the country south of Madras; **Telugu** that of the people inhabiting a district which may be described as a semicircle of which the centre is Rajahmundry (Rájamáhendripuram), and thence to Madras, the radius; **Kanarese** (Kannadam) is the language of Mysore (Maisúr) and Kanara (Kannada), and **Uriya** that of Orissa. **Malayálam** is spoken on the western coast from Mangalúr to Trivandram (Tiruvananthapuram); the aboriginal tribes speak their own distinct tongues.

29. *Industry.*—More than two-thirds of the people of India are engaged in **agriculture**. The preparation of indigo and opium, the manufacture of cotton and silk fabrics, coarse woollens, jewellery, and ornamental ware, salt, oil, brass utensils, coarse iron implements, and pottery are almost the only manufactures in the country.

[*Internal Communication.*—Up till a very recent period the generality of the roads in India were little more than mere tracks, scarcely suited to the transport of wheeled carriages. Travellers of the wealthier class journeyed at vast expense, either in palanquins carried on men's shoulders at the rate of four miles an hour, or on ponies at the rate of three, while the servants with baggage and tents followed on foot. Poor men travelled on foot by day, and rested under trees by night. Public coaches, waggons, and boats did not exist. There were no bridges, no railways, no telegraphs, no postal system. The roads, such as they were, were infested with wild beasts and robbers; and too often fever or cholera overtook the traveller, and suddenly ended his journey. Rest-houses were few and far between. Letters, money, jewels, and other valuables were despatched by single messengers, who trusted for security by the way solely to the poverty of their appearance. Merchandise was transferred in carts, or on the backs of camels, bullocks, or ponies, for distances of five hundred or even a thousand miles. Under British government most of these drawbacks have passed away; good roads exist now between all the large towns, streams have been bridged, canals cut, railways constructed, and rivers made navigable; and several millions sterling are annually expended in improving the existing means of communication and constructing public works. Besides, steamers ply all round the coast from Rangoon (Rangún) to Karáchi; telegraph wires traverse the land in all directions; letters and packages can be sent at small expense to all parts of the country; and though much remains to be done in the way of extension and improvement, at the present rate of progress it will ere long be possible to travel from any one part of India to any other at least as cheaply and with as much ease, speed, and security as is now possible in the most advanced countries of the world.]

30. *Government.*—The control of the whole civil and military government of the British territory in India is vested in a Governor-General and Councillors, styled "**The Governor-General in Council, or Government of India.**" In the Presidencies and Provinces the Government is administered by Governors, Lieutenant-Governors, or Chief-Commissioners, subject in all respects to the orders of the Governor-General in Council. For administrative purposes the whole is divided into shires called **Zilas** or **Collectorates,** each of which is placed under the charge of a Government Officer and has its peculiar courts, while these are again divided into **Táluks** and subdivided into **Villages.** The numerous Native States are tributary to the British Government, but their respective rulers are, in most cases, supreme in their own dominions. Nepál and Bhután are the only states in India quite independent of British rule.

31. *Revenue.*—The revenue of British India has much increased of late years, and is still increasing. It is derived chiefly from land, opium, salt, customs, stamps, and excise. The gross revenue is about **fifty crores of rupees.**

[The land revenue yields twenty crores; opium, nine; salt, six; customs, excise, and stamps, seven; and other smaller items yield the remaining eight. The receipts generally exceed the expenditure. About five crores are spent annually on the construction of public works. The cost of collecting the land revenue is about two and a half crores; and the maintenance of the army costs twelve.]

32. *Political Divisions.*—The present great divisions are the **Madras Presidency,** capital *Madras;* the **Nizám's Dominions,** capital *Hyderabad* (Haidarábád); the **Assigned Districts,** capital *Akola;* the **Central Provinces,** capital *Nágpur;* the **Lower Provinces of Bengal,** capital *Calcutta;* **Assam** (Asam); **British Burma** (Barma), capital *Rangoon* (Rangún); **Bhután**; **Nepál,** capital *Katmándu;* the **North-West Provinces** and **Oudh,** capital *Allahábád;* **Central India,** or **Málwa**; the **Bombay Presidency,** capital *Bombay;* **Rájputána**; the **Panjáb,** capital *Láhor;* **Kashmír,** and the **Hill States.** To these must be added **Ceylon,** capital *Colombo.*

---

## THE MADRAS PRESIDENCY.

33. *Position.*—The Madras Presidency extends from Cape Comorin in lat. 8° 4′ N., to the northern extremity of Ganjám in lat. 20° 18′, and from E. long. 74° 9′ to 85° 15′.

34. *Boundaries, Dimensions.*—Bombay, the Nizám's Dominions, and Bengal bound it on the north, and its eastern, southern, and western sides are formed by an almost unbroken line of sea-coast nearly 2000 miles in length. Its greatest length is about 950 miles, and the greatest breadth about half that distance. It has an area of about 150,000 square miles.

35. *Climate.*—It has been already stated that Southern India is exposed to the full force of the north-east and south-west monsoons, and that more rain falls on the west coast than on the east. Some districts, as Coimbatore (Kóyambuttúr), share in the rain brought by both monsoons, while those which are far removed and separated from the sea by ranges of mountains, as Bellary (Ballári), get the least rain of all. **The Madras Presidency is the hottest part of India.**

[On the Malabar Coast, where the atmosphere is moist, the mean temperature is 78°, being seldom lower than 68° or higher than 88°; on the Coromandel Coast the average temperature is 84°, rising occasionally during the hot season to more than 100°. Throughout the Presidency, previous to the rains, the country has a very arid appearance, and in the Carnatic (Karnátik), that is, in the districts of Nellore (Nellúr), North and South Arcot (Arkát), Madras, Trichinopoly (Tiruchirápalli), Tanjore (Tanjávúr), Madura (Mathurai), and Tinnevelli (Tirunelvéli), in May, June, and July the grass is scorched up, all vegetation droops, trees shed their leaves, rivers and tanks become dry, animal life suffers, and strong hot winds laden with dust blow from the west. In August the heat, dust, and glare are somewhat moderated by occasional heavy thunderstorms; but a few days after the great rains, which fall about the end of October, the surface of the whole country becomes changed as if by magic from a naked arid expanse to a sheet of the most varied and luxuriant verdure; trees put forth fresh foliage and all nature is refreshed. Along the coast the sea-breezes, which set in shortly after noon almost throughout the year, do much to moderate the temperature.]

36. *Natural Productions.*—MINERAL PRODUCTS.—**Iron ore** occurs in several places, but in abundance in South Arcot (Árkát) and Malabar (Malaiyálam); **copper ore** in Nellore (Nellúr) and the Eastern Gháts; **magnesia** in Salem (Sélam); and **salt** is obtained from the sea by evaporation. VEGETABLE PRODUCTS.—**Rice** is grown throughout the Presidency, but especially in the alluvial grounds of Godávari, Krishna, Tanjore (Tanjávúr), Malabar (Malaiyálam), and Kanara (Kannada). **Maize, millet,** and **rági** are also everywhere largely cultivated; so also are **oil-seeds, tobacco,** and **sugar-cane.** Along the coasts and in sandy tracts **cocoanut** and other **palms** are extensively grown. **Cotton** is grown mostly in Cuddapah (Kadapa), Kurnool (Karnúl), Bellary (Ballári), and Tinnevelly (Tirunelvéli); **indigo** in Cuddapah (Kadapa) and Nellore (Nellúr); **coffee** on the Palnais, Shervaráis, and Nílgiris; **tea** and **chinchona** on the Nílgiris; and **pepper** and **cardamoms** on the western coast. Most of the hills are covered with forests producing a great variety of **drugs, dyes,** and **gums,** and many sorts of the finest **timber,** such as teak, sál, sísu, blackwood, and sandalwood; while oranges, limes, mangoes, plantains, pineapples, melons, and other Indian **fruits** are produced almost everywhere in great abundance. ANIMALS.—**Elephants, tigers, cheetahs, bears,** and **bison** frequent the forests; many sorts of **deer** are met with in all parts; **monkeys** and **jackals** are numerous in the cultivated country and in towns; **lizards, snakes,** and other reptiles are found in all places; as are **crows, kites,** and other birds, and **mosquitoes** and other insects. **Fish** is plentiful in all the rivers and along the coast; **oxen** are

numerous, and are kept for draught purposes; **buffaloes, sheep, goats,** and **poultry** are everywhere common.

37. *People.*—The population exceeds 31,000,000.

Wayside Temple, Madras Presidency.

38. *Religion.*—The great bulk of the people profess the **Hindu** religion; Bráhmans are numerous, and congregate especially in the district of Tanjore

(Tanjávúr). Bráhman temples, some of vast size, may be seen in every town in Southern India, but those of Tanjore (Tanjávúr), Sríirangam, and Madura (Mathurai) are especially renowned. **Muhammadans**, 1,800,000 in number, are fewer in proportion to the population that in the north of India. All Europeans and their descendants, and many natives, especially in Tinnevelly (Tirunelvéli), are **Christians**; and there are many **Jews** on the Malabar coast.

39. *Language.*—The principal languages are **Tamil**, **Telugu**, **Malayálam**, and **Kanarese**.*

[*Telugu* is spoken by 11,600,000 people in the districts from Ganjám on the north, to Pulicat on the south, and to Bellary (Ballári) on the west; *Tamil* is used in the south part of North Arcot (Árkát) and throughout the districts of Madras, South Arcot (Árkát), Trichinopoly (Tiruchirápalli), Tanjore (Tanjávúr), Madura (Mathurai), Tinnevelly (Tirunelvéli), Salem (Sélam), and Coimbatore (Kóyambuttúr), by 14,700,000 people; *Kanarese* (Kannadam) in Bellary (Ballári) and Kanara (Kannada) by 1,700,000; *Malayálam* in Malabar (Malaiyálam) and North Travancore (Tiruvángkúr) by 2,300,000; *Uriya* in the north of Ganjám by about 500,000; *Tulu* in part of South Kanara (Kannada) by 30,000; and *Hindustáni* and *English* throughout.]

40. *Education.*—Education in Madras, as everywhere else in India, is rapidly extending. Public instruction is under a Director.

[The number of schools under Government inspection is about 10,000, and the number of pupils attending about 300,000. There are besides hundreds of indigenous village schools under native management. The annual expenditure by Government is about sixteen lakhs of rupees.]

---

* These languages are all very closely allied, and are classed as "*Dravidian*," *from Drávida* or *Drávira*, the ancient name of a tract of country nearly conterminous with that now occupied by the Tamil race.

41. *Industry.*—Two-thirds of the population are engaged in **agriculture.** The only important manufactures are cotton cloths, sugar, indigo, brass vessels, and pottery.

[Coarse earthen vessels are made in almost every town and village, and weaving and dyeing are carried on to a trifling extent in almost every town. In some few places, as Trichinopoly (Tiruchirápalli), small trades, such as the manufacture of jewellery and other articles of taste, are carried on.]

42. *Internal Communication.*—The rivers of this Presidency being almost unnavigable, communication is held by means of **roads, canals,** and **railways.** Good roads connect all the large towns.

[Many of the roads are kept in excellent order and lined on both sides with avenues of trees, affording pleasure to the eye of the traveller and protection from the rays of the sun. Rest-houses are erected, either by the Government or the bounty of individuals, at intervals of every ten or twelve miles; bridges are constructed over deep streams, and, except from occasional accident caused by the heavy periodical rains, communication between most parts of the Presidency is easy and uninterrupted. Many of the rest-houses built and endowed by private individuals are spacious structures adapted to accommodate travellers of all castes and creeds. In all those erected by Hindus, one portion is set apart for Sudras and another for Bráhmans; in some there is also accommodation provided for others of a different creed; rooms are provided free of expense, and in the best of them cooks and other servants also. In some, the traveller obtains without any payment, curry, rice, ghee, and fruit. The great man's rice may be a little whiter than the poor man's, the ghee a little fresher, the fruit more choice, and the quantity greater, but all receive as much as, or more than, suffices; and should the well-supplied traveller decline to receive the gift, he would be considered a churl, and his refusal as an insult to the memory of the founder. In this way the Hindus, who are often most unjustly considered an inhospitable people, show much hospitality to strangers.]

43. Canals.—The canals of this Presidency are not

on such a large scale as those of some other parts of India. They are constructed chiefly in connection with the systems of irrigation in the Godávari, the Krishna, and the Káveri deltas.

[In each of these localities there is a perfect network of irrigating canals, the larger channels serving also as lines of navigation. Along the eastern coast, a continuous series of salt-water canals run through the districts of South Arcot (Árkát), Chingleput (Chengalpat), Madras, and Nellore (Nellúr). Great traffic is carried on in these in fish, firewood, chillies, salt, and shells for lime. On the western coast the shallow parts of the Cochin (Kochi) "backwater" have been deepened, and an excellent channel of communication exists for nearly two hundred miles, along which the rich products of Travancore (Tiruvángkúr) and Malabar (Malaiyálam) are transported to the sea.]

[*Railways.*—The *Madras Railway* runs S.W. by Salem (Sélam) and Coimbatore (Kóyambuttúr) to Beypore (Bépur) on the Malabar Coast, a distance of 406 miles. At Coimbatore (Kóyambuttúr) a short line branches off to Metapálliam near the foot of the Nílgiris; at Jalárapet, 132 miles from Madras, another line diverges to Bangalore (Bengulúr), a distance of 84 miles; and at Árkonam, 42 miles from Madras, another line branches north-west by Cuddapah (Kadapa) and Gutti to the town of Raichúr (Ráyachúru) in the fertile valley of the Raichúr Duáb, where it joins the line from Bombay. The *South Indian Railway* extends from Madras to Tanjore (Tanjávúr) 217 miles; from Trichinopoly (Tiruchirápalli) to Tuticorin (Túttukudi) and Tinnevelly (Tirunelvéli); and from Negapatam (Nágapatnam) to Erode (Iródu) Junction on the Madras line. A light line of railway, running from Chingleput (Chengalpat) to Conjeveram (Kánchípuram), and thence to Árkonam, serves to connect the *Madras* and the *South Indian Railway* systems.]

44. *Commerce.*—IMPORTS—**Cotton, woollen,** and **hardware goods, manufactured metals, books, wines, spirits, timber,** and **horses.** EXPORTS—**Cotton, sugar, coffee, indigo, rice, hides, jaggery, cocoanut-oil, oil-seeds, cardamoms, ginger,** and **pepper.** The greater

part of the trade is with the United Kingdom; the rest with Bombay, Calcutta, and other Indian ports, Ceylon, Singapore, France, Mauritius, and Australia.

45. *Ports.*—These, though numerous, are by no means well adapted for commerce, for, notwithstanding the great length of coast-line, harbours are few. The chief ports are **Bimlipatam** (Bhímunipattanam), **Vizagapatam** (Vishákhapattanam), **Cocanada** (Kákináda), **Masulipatam** (Machílípatnam), **Madras, Pondicherry** (Puduchéri), **Negapatam** (Nágapatnam), **Tuticorin** (Túttukudi), **Cochin** (Kochi), **Calicut** (Kalikód), **Tellicheri** (Talachéri), **Cannanore** (Kannúr), and **Mangalore** (Mangalúr).

46. *Government.*—The twenty-one districts are subject to the direct control of "**The Governor of Madras in Council**;" Jaipúr is under the superintendence of the Governor's Agent at Vizagapatam (Vishákhapattanam); Pudukotta (Pudukottai) under that of the Collector of Tanjore (Tanjávúr); and Cochin (Kochi) and Travancore (Tiruvángkúr) are ruled by their respective rájas, subject to the interference of the British Government.

47. *Revenue.*—The revenue is derived from land, salt, customs, stamps, excise, and sundry other sources, as forests and fisheries. It is increasing, and amounts annually to rather more than **seven crores of rupees.**

48. *Condition of the People.*—The masses of the people are, in common with those of other parts of India, very poor; but considering their simple habits and the nature of the climate, they are far from being

in that uncomfortable state in which the people of colder countries would be on the same poor incomes.

[Except in Bellary (Ballári) and the neighbouring districts, where flat-roofed houses built of stone are the rule, the agricultural labouring population live chiefly in mud-walled huts with thatched roofs. Generally the hut contains but one room, having one aperture only, which serves for doorway and window. A plaited cocoanut leaf, or a mat of woven strips of the leaf of the palmyra, which serves for a bed, and a few earthen vessels, compose the scanty furniture. Rági forms the chief article of food of the poorer classes, and a narrow strip of cloth wrapped round the loins is all the clothing they wear. Men, women, and children work in the fields, and the earnings of a whole family amount, if the work be continuous, to perhaps from six to eight rupees a month, but the rate of labourers' wages varies much in different districts. In many localities agricultural work is paid for wholly in kind, in others partly in kind and partly in coin. In the neighbourhood of towns wages are higher, the mud hut is displaced by one of bricks and mortar, the earthen pots give way to vessels of brass, rice takes the place of rági, and the dress of the people, though by no means full, is much less scanty. In towns the great bulk of the inhabitants are either shopkeepers or artificers, and their condition is somewhat better.]

49. *Divisions.*—The Presidency is divided into twenty-one districts, viz., **Ganjám**, **Vizagapatam** (Vishákhapattanam), **Godávari**, **Krishna**, **Nellore** (Nellúr), **Madras**, **Chingleput** (Chengalpat), **North Arcot** (A'rkát), **South Arcot** (A'rkát), **Trichinopoly** (Tiruchirápalli), **Tanjore** (Tanjávúr), **Madura** (Mathurai), **Tinnevelly** (Tirunelvéli), **Cuddapah** (Kadapa), **Kurnool** (Karnúl), **Bellary** (Ballári), **Salem** (Sélam), **Coimbatore** (Kóyambuttúr), the **Nílgiris**, **Malabar** (Malaiyálam), and **South Kanara** (Kannada). Locally and politically connected with it are **Jaipur**, **Puducotta** (Pudukottai), **Cochin** (Kochi), and **Travancore** (Tiruvángkúr).

50. **Madras**, situated on the east coast in 80° 16′ E

long. and 13° 5′ N. lat., is the seat of the British Government in Southern India, and an extensive place with about 400,000 inhabitants. It is connected with the western coast by rail, and another railway joins it with Bombay. Steamers leave regularly for England with goods and passengers, and for every important maritime town between Rangoon (Rangún) and Karáchi. Madras is badly situated for commerce, as ships cannot approach very near the shore. Lately, however, the place has been greatly improved as a port by the construction of an artificial harbour, in which vessels can anchor in smooth water. Madras has considerable trade with Britain, Calcutta, and Burma (Barma).

[The town of Madras and its suburbs form at once the Municipality and the little District of Madras, which is only twenty-seven square miles in area. A piece of land having been obtained from a native prince in 1639, the English constructed Fort St. George, and thus formed the nucleus of the present city and its suburbs. In 1702 the fort was besieged unsuccessfully by Dáud Khán, but in 1744 it fell before the attacks of the French under Labourdonnais, and was retained by them until the peace of 1749, when it was restored. Again in 1758 it was assailed by the French under Count Lally, but this time unsuccessfully. In 1769 it was threatened by Haidar Ali, and since then it has never been molested by an enemy.]

51. **Ganjám.**—This, the most northern district of the Presidency, has an undulating surface rising gradually towards the west into hills which contain the sources of all the streams that water the district. Rice, sugar-cane, maize, millet, rági, and oil-seeds are produced in the plains in abundance; and the hills furnish wax, gums, drugs, dyes, and timber. Cotton is also grown in considerable quantity. The coast of Gánjam is bold and rocky; but, like other parts of the Madras Presi-

dency on the eastern coast, it possesses no harbour for shipping.

[The salt-water Lake of Chilká between this district and Purí on the north is forty-two miles long, fifteen broad, and extremely shallow, its depth nowhere exceeding six feet. It contains several inhabited islands, and is separated from the sea only by a narrow strip of sand. Much excellent salt is made from it by evaporation.]

*Towns.*—**Chicacole** (Srîkákulam) is the chief town of the district. **Berhampore** (Barahampuram) is a large town situated in the midst of a cultivated plain.

[This part of India was anciently called *Kalinga*, a name still retained in that of the modern town of Kalingapatnam.]

52. **Vizagapatam** (Vishákhapattanam).—The district of Vizagapatam (Vishákhapattanam) resembles Ganjám in many respects. It has, like it, a bold rocky coast, and an undulating surface gradually rising towards the west. The streams are small and dry during a part of the year, and the productions are similar to those of Ganjám. Cotton goods are manufactured to some extent, and iron is found in many parts. Much of the district is composed of Zamindaries, and the great estates are almost entirely under the control of their Zamindars or Chiefs. To the west of the district lies the large State of Jaipúr, a rugged country about 10,000 square miles in extent, destitute of roads and other evidences of civilisation, and inhabited by Khond tribes who are governed in feudal fashion by their own Zamindars, all of whom are subject to British authority.

[The Khonds dwell in villages in the forests. Each village consists of two streets composed of a double row of huts, well built of timber, daubed with mud, and having thatched roofs. In some cases the villages are stockaded. The Khonds are a warlike people; they carry bows, arrows, and axes, but wear no head-dress, and in fact very little clothing. Their chief occupation is hunting. Their agriculture is of the rudest kind; a piece of forest is marked out and the ground cleared by fire; small holes are then chopped in the earth by axes, and the seeds are dropped in. After a year or two this is deserted for a fresh piece of forest land cleared in like manner. Police, schools, and fairs have of late years been established in the Khond districts, and much good has resulted.]

*Towns.* — **Vizagapatam** (VisháKhapattanam), the chief town of the district, is famed for its elegant elk-horn, ivory, and sandalwood boxes. Near it is **Waltair** (Waltér). **Vizianagram** (Vijayanagaram = the Town of Victory) is the chief town in the large Zamindari of the same name. **Bimlipatam** (Bhímunipattanam = Bhíma's Town) is a maritime town of rising importance.

53. **Godávari.**—In common with the two former districts, that of Godávari is very fertile, and the delta of the river one of the most luxuriant tracts in all India. Near the coast the land is low and level, but in the north and north-eastern parts hilly. A dam or anicut built near Dhauléshwaram prevents the Godávari from running uselessly to the sea, so that an unceasing supply of water is available for purposes of irrigation. In the course of the river numerous small islands (lankas) are formed, of surprising fertility; these are carefully cultivated, chiefly with tobacco, which grows in such soil to great perfection. The low grounds are subject to inundations. In the bed of the Godávari, agates, cornelians, and chalcedony are found. Rice, maize, millet, rágí, sugar-cane, cotton, indigo, and tobacco are largely produced and exported. The navigation along the coast is rendered dangerous by shoals and sandbanks.

**[Partly in the Godávari district and partly in that of the Krishna, lies the fresh-water lake of Kolair (Kollér). It is of oval shape, and in the dry season is about twenty-five miles in length and ten in breadth. It contains many islands. During the rains its length, breadth, and depth are greatly increased, and the islands disappear.]**

*Towns.*—**Rajahmundry** (Rájamáhendripuram) is the chief town. **Coringa** (Kórangi), **Cocanada** (Kákináda), and **Narasápur** are small ports. **Ellúr** has carpet manufactures.

54. **Krishna.**—Near the coast, and for a distance of about fifty miles inland, the surface of this district is depressed, and in some parts below the level of the sea; westward it rises into hills of considerable height, attaining their greatest elevation near the town of Kondapalli. The navigation near the coast is rendered dangerous by shoals which extend five or six miles out to sea. As in the case of the Godávari, the Krishna is by means of anicuts and canals made to irrigate and fertilise a large portion of the district, instead of being allowed to cast its waters uselessly into the ocean. The vegetable productions are chiefly rice, oil-seeds, and tobacco. Salt is made along the coast.

*Towns.*—**Masulipatam** (Machílípatnam) or **Bandar**, the chief town of the district, is situated on a branch of the river Krishna. Printed cotton goods are here manufactured; so also is snuff. In a dreadful hurricane in November 1864, a huge wave of the sea suddenly rose at midnight and was driven inland like a wall several feet high. This town and the villages for twelve miles round were utterly destroyed, while thousands of animals and 30,000 people were drowned. **Beswára** (Bejwáda) stands on the trunk road from Madras to Calcutta. **Guntúr** (Gundúr) is another town of local importance.

[The districts of Ganjám, Vizagapatam (Visbákhapattanam), Godávari, and Krishna are the present divisions of the country formerly called the "Northern Circars." They were obtained by the French in 1753, and remained in their possession till 1759, when they were seized by Clive, and subsequently formally ceded to the British by the Emperor of Delhi.]

55. **Nellore** (Nellúr).—Nellore (Nellúr) is in general flat, and near the coast low and sandy. The soil is productive only in the vicinity of tanks and rivers, consequently more than half the district lies waste, and much of it is overrun with jungle. Along the coast cocoanut and palmyra palms grow in great luxuriance. Rice, rági, oil-seeds, cotton, indigo, and tobacco are all grown to some extent; salt is made in great quantity; and this district has a fine breed of bullocks in much request for draught purposes all over Southern India.

[A wild race of people, called Yénádis, inhabit the jungles along the sea-shore. In habits, religion, and language they are quite distinct from their neighbours; they are short in stature, and of black complexion; capable of enduring great fatigue, and remarkably faithful and honest. They live on roots, fruits, leaves, fish, and rats, and have no industrial pursuits.]

*Towns.*—**Nellore** (Nellúr), the chief town of the district, and **Ongól** (Vángavól), are the only places of importance.

[Nellore (Nellúr) was transferred to the British by the Nawáb of Arcot (Árkát) in 1801.]

56. **Chingleput** (Chengalpat).—This district is everywhere flat, with the exception of a few rocky isolated hills; the soil where well watered is fertile. The streams are few, and dry in the hot season, and the cultivation is dependent on the supply of water stored up in tanks, of which there are great numbers, some of vast extent. Rice, rági, sugar-cane, and betel are chiefly cultivated.

*Towns.*—**Chingleput** (Chengalpat) was till lately the chief civil station in the District; **Saidápet** (Saidápéttai), near Madras, is now the headquarters of the Collector. **Conjeveram** (Kánchípuram) contains several large pagodas, and was formerly the capital of the ancient Hindu kingdom of Chóla. **Sadras** (Sathurangapattanam) and **Pulicat** were both in the possession of the Dutch once, and were made over to the British in 1819. **Mahábalipuram,*** or the *Seven Pagodas*, near Sadras (Sathurangapattanam), is a collection of extraordinary monolithic temples covered with sculptures. Tradition points to it as the site of a great city, part of which is now covered by the sea.

---

* "City of the Great Bali." According to the Puránas, Bali was a very powerful king, and increased in wealth and mightiness so much that he grew arrogant and set himself up as a god. Vishnu resolving to punish him, one day appeared to Bali as Vámana, a Bráhman dwarf, and requested as much ground for his maintenance as he could measure in three strides. Bali readily assented, and to ratify the donation poured water on Vámana's hand. Instantly the dwarf began to assume giant dimensions, and increased in stature until his head reached the stars; then with one stride he measured the whole earth, with another all heaven, and for the third he placed his foot crushingly on the terrified Bali's head and sent him to reign in the realms below.

57. **North Arcot** (Árkát).—This district and the neighbouring one of South Arcot (Árkát) form part of the slope from the Eastern Gháts to the sea. The eastern portion is low and level, but that to the west is broken into isolated ranges of hills of variable height. The Nagari Hills, in the north of the district, form a prominent object far out at sea. The soil of North Arcot (Árkát) is moderately fertile; there are numerous streams, and the cultivation is aided by artificial lakes, tanks, and wells; cotton, rice, and rági are chiefly grown.

*Towns.*—**Arcot** (Árkát), on the Pálár, was the capital of the Carnatic (Karnátik) under Muhammadan rule, and its fort, now in ruins, was gallantly held by Clive with a handful of troops when besieged by Chanda Sáhib's forces in 1751. **Vellore** (Velúr), also on the Pálár, has a fort in which State prisoners have from time to time been confined; it has also a large Sivite pagoda. **Chittúr** is the chief town in the district. **Arni** (Árani) and **Vándivásh** (Vándivásam) are both associated with the wars in the Carnatic (Karnátik). **Tripeti** (Tiruppati), in the north of the district, is a place of pilgrimage. **Sátgiri** (Sátthukadi) is noted for the superior oranges grown in its neighbourhood.

[The country stretching from the Gandigáma to Cape Comorin was in ancient times divided between the kingdoms of Karnáta, Chóla, and Pándya; but when the Europeans first came to India, in the sixteenth century, these kingdoms had disappeared, and the greater part of the country acknowledged the authority of the Hindu king of Bijianagar. In 1565 this kingdom was destroyed by the Dakhan Musalmáns, when the last rája being slain, his brother fled to Chandragiri, where he established himself as king of the country between the Eastern Gháts and the sea. In 1639 one of the successors of this prince granted to the E. I.

Company a strip of land on which they built Fort St. George, but in 1646 the Musalmán kings of Bijapur and Golkonda conquered him and divided his dominions between them, the land around and to the north of Madras falling to the latter. In 1687 these two kingdoms were subverted by Aurangzéb, whose authority was acknowledged as far as Cape Comorin by the close of the century. Thenceforward the modern Carnatic (Karnátik) was governed by Mughal officers called Nawábs, until it was forced upon the English to take charge of the country. The first Nawáb seems to have been Dáud Khán, the same that unsuccessfully besieged Madras in 1702. In 1710 Sa'ádat Ullah was appointed Nawáb of Arcot (Árkát), and in consequence of the distractions of the Mughal empire, he not only retained his office as long as he lived, but on his death in 1732 was able to transmit it to his nephew Dust Ali, who being killed in battle by the Máráthas in 1740, was succeeded by his eldest son, Sufdar Ali. The latter was murdered at Vellore (Vélur) in 1742, and his infant son, Saiyid Muhammad, was proclaimed. The next year Nizám-ul-Mulk took charge of the country, and placed both it and the young prince under Anwar-ud-dín as guardian. In 1744 Saiyid Muhammad was murdered, and the first family of Nawábs came to an end. The guardian, Anwar-ud-dín, was then appointed Nawáb by the Nizám, but being slain in battle at Ambúr in 1749, and his eldest son, Mafúz Khán, being taken prisoner, the whole country except Trichinopoly (Tiruchirápalli) fell into the hands of Chanda Sáhib, the son-in-law of Dust Ali. Muhammad Ali, the second son of Anwar-ud-dín, held out in Trichinopoly (Tiruchirápalli) until 1751, when, being reduced to the last extemity, he offered to give up the city and renounce all claim to the Carnatic (Karnátik) if the Nizám would give him a command in any other part of his territory; but just at that crisis, before he received an answer to his proposal, the English, with Clive, came to his assistance, and after a struggle of several years succeeded in getting the whole country into their hands by 1761. They then made Muhammad Ali Nawáb of Arcot (Árkát); but he was so little able to protect himself that he removed his family from his own capital to Madras, where they and their descendants have ever since continued to reside. He not only trusted to the English for the defence of himself and his country, but was obliged constantly to call in their aid to enforce the payment of his revenues. He took the name of Wálla Jáh on his investiture as Nawáb, and on his death in 1795, the English made his eldest son, Umdat-ul-Umará, his successor. Umdat-ul-Umará died in 1801, and the English then found it necessary to assume full charge of the country. The title, with a very large revenue, was, however, given to his brother Azam-ud-Daulah. Several princes subsequently bore it, but it ceased in 1855. The lineal male representative of the last Nawábs of the Carnatic (Karnátik) is now styled "Prince of Arcot."]

58. **South Arcot** (Árkát).—In physical aspect this district does not differ from North Arcot (Árkát). The productions are also similar. By means of an anicut and artificial channels, the Coleroon (Kolladam) is made to irrigate and fertilise a large tract of country.

*Towns.*—**Cuddalore** (Gúdalúr) is the chief town in the district. Near it is **Fort St. David**, once the capital of the British possessions on the Coromandel Coast. **Jinji** is a hill fortress, much contended for in the wars of the Carnatic (Karnátik).

59. **Trichinopoly** (Tiruchirápalli).—The districts of Trichinopoly (Tiruchirápalli), Tanjore (Tanjávúr), the eastern part of Madura (Mathurai), and Tinnevelly (Tirunelvéli), are all of the same general flat character, diversified only by scattered high granite rocks. They are all well watered, and produce rice, rági, millet, tobacco, and cotton.

*Towns.*—**Trichinopoly** (Tiruchirápalli) is situated on the Káveri. The natives are famed for their skill in the manufacture of jewellery, hardware, saddlery, and cheroots. On a remarkable rock, six hundred feet above the plain and in the town, there is a very large pagoda, very much resorted to by devotees. Near this town is **Srírangam** (=the divine Vishnu), a famous place of pilgrimage, on an island of the same name formed by the Káveri. At its western extremity is a celebrated pagoda enclosed by seven square walls, each twenty-five feet high and four feet thick, and the outer one four miles in circuit. Each enclosure has four gates facing the cardinal points, and surmounted by lofty towers. Within the area, besides the pagoda itself, are streets containing shops, choultries, small temples, and the houses of the Bráhmans.

**[Trichinopoly (Tiruchirápalli) was included in the Carnatic (Karnátik), and consequently came under British control in 1801.]**

Entrance to the Temple, Srírangam.

60. **Tanjore** (Tanjávúr).—Tanjore (Tanjávúr) forms a part of the delta of the Káveri, and is consequently altogether flat. An elaborate and costly irrigation system of anicuts, embankments, and canals, connected with the Káveri and Coleroon (Kolladam), furnishes a never-failing supply of water to almost the whole of the district. Tanjore (Tanjávúr) is consequently not surpassed in productiveness by any district in India. It yields more than one-tenth of the entire revenue of the Presidency.

*Towns.*—**Tanjore** (Tanjávúr, the ancient *Tanjánagaram* = City of Refuge) has a pagoda considered the finest in India; in front of it is a colossal figure of a bull in black granite. Tanjore (Tanjávúr) carries on much trade, and has manufactures of silk, muslins, and cottons. **Combaconum** (Kumbhakonam) is a large town inhabited by many Bráhmans. It has much weaving and working in brass, and a provincial college. **Nagore** (Nágúr), **Negapatam** (Nágapatnam), and **Tranquebar** (Tarangambádi) are all on the coast. Tranquebar (Tarangambádi) was purchased by the British from the Danes in 1845. **Máyavaram**, a large town, is a place of pilgrimage. **Shiali** (Sirkáli) is another large town.

[In 1678, Venkáji, brother of the famous Siváji, obtained possession of this province, and his descendants exercised uncontrolled authority over it till about the year 1770, when a claim to tribute was made to Tuljáji, the rája, by Muhammad Ali, Nawáb of Árkát. The Madras Government supported the claim, and in 1772 the English captured the fort, deposed Tuljáji, and placed the management of the country in the hands of the Nawáb. The Home authorities disapproved of these proceedings, directed the reinstatement of the deposed prince, and entered into a treaty with him by which Tuljáji agreed to pay a large annual subsidy to the British. Shortly afterwards Tuljáji died, and Amír Singh, his half-brother, and Sarabhóji, Tuljáji's adopted son, were claimants for the throne. Amír Singh was chosen, but governing badly, he was set aside in favour of Sarabhóji in 1798. The following year Sarabhóji gave up the management of the country to the British on condition of receiving rather more than one-fifth of the revenue. Sarabhóji died in 1832, and was succeeded by his son Siváji, who dying in 1855 without leaving a son by birth or adoption, the title of the rájas of Tanjore (Tanjávúr) was declared to be extinct.]

---

61. **Puducotta** (Pudukóttai).—Southward of Tanjore (Tanjávúr), and subject to its supervision, is the State of Puducotta (Pudukóttai), under the government of a rája called "the Tondamán." Much of it is covered with dense jungle. The principal place is **Puducotta** (Pudukóttai).

---

62. **Madura** (Mathurai).—Madura (Mathurai) may be considered as an inclined plane, rugged and elevated towards the west and north-west, whence it slopes more or less gradually till it sinks into low plains, a few feet above the sea. It is watered by many streams, the largest of which is the Vaigai. The Palnai hills in the north-west of the district have a general elevation of about 7000 feet above the sea. Rugged heights, deep ravines, fertile valleys, dense forests, winding streams, and scattered villages, all blending together, make the scenery of these hills in the highest degree picturesque. They are used by Europeans as a sanitarium. The coast of this district being beset with dangerous rocks, it cannot be visited by vessels of any considerable burthen. Adam's Bridge (or Ráma's Bridge, as the

Hindus call it), a narrow ridge of sand and rocks, mostly dry, nearly connects the extremity of Rámnád with Ceylon; one end of this ridge joins the island of Mannár, beween which and Ceylon is a narrow passage navigable for vessels of moderate size, and the other joins the much celebrated island of Ráméswaram, between which and the mainland is the Pámban passage, navigable for ships of 500 tons. Madura (Mathurai) produces timber, cotton, sugar-cane, tobacco, and grain. The manufactures are white and coloured cotton cloths.

*Towns.*—**Madura** (Mathurai) is the principal place in the district. It was the capital of the Hindu kingdom of Madura (Mathurai), and was formerly the seat of learning in Southern India. It is of great antiquity. Weaving is carried on to a considerable extent, the cloths being much prized for their brilliant colours; there is also much working in brass. **Sivaganga** is the chief town of the extensive Zamindari of the same name; **Dindigal** (Dindukkal) is noted for its cheroots; and **Rámnád** is the chief place of the large Zamindari of that name.

63. **Tinnevelly** (Tirunelvéli). — The coast of this district is everywhere low and level, but rendered very dangerous by rocks and shoals. Inland the ground rises into a plain elevated about 200 feet above the sea, and terminated on the Travancore (Tiruvángkúr) side by the Western Gháts, which have, in that part, a general elevation of 4000 feet. Many streams rising in the mountains on the west, and filled by the rains brought by the south-west monsoon, water the district.

The chief is the Támbiraparuni (Sanskrit *támra parní* = the Copper-coloured). The soil is for the most part poor, and of a deep red or rusty colour; much cotton, however, is grown besides the usual crops. In the south-east parts, especially where the soil is of a sandy nature, vast numbers of palmyra trees grow. The cultivation of these, the drawing of toddy, and the manufacture of jaggery, form the chief occupations of the people.

**[It is in Tinnevelly (Tirunelvéli) that most converts from Hinduism to Christianity have been made. There are several villages and small towns, as Nazareth and Christianagram, almost exclusively occupied by them. In the neighbourhood of Tuticorin (Túttukudi) the pearl-fishery is carried on, but the pearls are neither so plentiful nor so valuable as those of Aripo on the opposite coast of Ceylon. Many hundreds of coolies annually leave Tinnevelly (Tirunelvéli) to work on the coffee estates in Ceylon, returning to their homes, when the coffee has been gathered in, with the money they have saved.]**

*Towns.*—**Palamcottah** (Pálaiyangkottai) is the chief civil station. It is near **Tinnevelly** (Tirunelvéli), the largest town in the district. **Tuticorin** (Túttukudi), on the coast, has pearl-fisheries in the neighbourhood, and some shipping trade, chiefly cotton. **Courtallum** (Kuttálam), a small town situated in a recess on the eastern side of the Western Gháts, 700 feet above the sea, is the resort, during the hot season, of invalids from the plains. The scenery around is exceedingly rich and varied, its beauty being increased by a series of waterfalls.

64. **Cuddapah** (Kadapa).—Cuddapah (Kadapa) is for the most part an open plain sloping eastward. A range of hills on the eastern side forms the boundary between Cuddapah (Kadapa) and Nellore (Nellúr).

Much of the soil, which contains soda, salt, and saltpetre, is of that description known as "black cotton ground;" it is remarkably fertile, and produces cotton in great abundance. Rice, rágí, tobacco, oil-seeds, indigo, and sugar-cane are extensively grown. The chief manufactures include cotton goods, coarse woollens, and the preparation of indigo.

*Towns.*—**Cuddapah** (Kadapa) is the principal town in the collectorate. **Ráyachúti** is the only other place of any importance in the district.

[Cuddapah (Kadapa), Bellary (Ballári), and the eastern part of Kurnool (Karnúl) were ceded to the British in 1800 by the Nizám, subsequently to the partition of Mysore (Maisúr), on Típu's downfall. They are known in history, therefore, as the "Ceded Districts."]

65. **Kurnool** (Karnúl).—Kurnool (Karnúl) is hilly almost throughout, the hills producing teak, blackwood, bamboo, and other valuable woods. The plains produce similar crops to those of Cuddapah (Kadapa), and the soil is for the most part black. The districts of Kurnool (Karnúl), Cuddapah (Kadapa), and Bellary (Ballári), being far inland, are beyond the full influence of the monsoons; consequently less rain falls in them than in any other part of Southern India, and the country is liable to great droughts. To prevent the evils arising from a scarcity of water, a system of irrigation works on a large scale has been constructed in the neighbourhood of Kurnool (Karnúl).

*Towns.*—**Kurnool** (Karnúl), on the Tungabhadra, is the chief town in the district. **Kambam** and **Nandiál** are towns of moderate size.

[The western half of Kurnool (Karnúl) was, until the year 1838, under the government of a Nawáb, a tributary of the British. While preparations were being made for restoring Sháh Suja to the throne of Afghánistán, it was discovered that a wide-spread Muhammadan conspiracy for the subversion of British power in India was formed, and that the Nawáb of Kurnool (Karnúl) was one of its originators. A small British force, therefore, entered the district, encountered the Nawáb at the head of his troops, and took him prisoner; and thenceforth his territory was annexed to the British dominions.]

66. **Bellary** (Ballári).—Bellary (Ballári) is an elevated tract rising towards the west to meet the Western Gháts, and towards the south to the elevated tableland of Mysore (Maisúr). The soil is very fertile, and produces much the same crops as that of Cuddapah (Kadapa), but perhaps less rain falls in this district than in any other in Southern India. Numerous huge, bare, dark-coloured granite rocks, which in several places start up abruptly from the ground, assuming the most fantastic shapes, form a distinguishing feature of this part of the country.

*Towns.*—**Bellary** (Ballári) is the chief town in the collectorate. **Gutti** and **Adwánni** are other places of local importance.

67. **Salem** (Sélam).—The Salem (Sélam) district is on a level with the Carnatic (Karnátik) in the south, but rises into a low tableland in the centre, and to a plateau on a level with Mysore (Maisúr) on the north. The Eastern Gháts traverse it on the east and north, and a rugged mountainous tract forms its western border. All the hills are inhabited and extensively cultivated. The Shervaráis near Salem (Sélam), about 5000 feet high, are much frequented by Europeans.

Erkád, the principal settlement, stands about 4500 feet above the sea. Most of the streams rising in this district flow south and south-west into the Káveri, which forms the western and south-western boundary; but some few flow north and north-east into the Pálár. Cotton, indigo, sugar-cane, tobacco, and the usual grains are the chief crops.

*Towns.*—**Salem** (Sélam) stands in a valley between the Shervaráis on the north and a small range of hills on the south. Other towns of importance are **Oossoor** (Osúr), **Tripatur** (Tiruppattúr), **Darumpuri**, and **Námkal** (Námakkal).

[Salem (Sélam) was ceded to the British in 1792 by Típu Sáhib.]

68. **Coimbatore** (Kóyambuttúr).—Coimbatore (Kóyambuttúr) is bounded on the north by an irregular mass of mountains separating it from Mysore (Maisúr); on the south by the Palnai and Ánamalai groups; and on the north-western side, beyond the deep gap of Pálghátchéri, by the Kundas and the Nílgiris. The Pálghát Valley, the only one which interrupts the chain of the Western Gháts from Comorin to Bombay, is about twenty miles wide, and near the centre has an elevation of about 1000 feet above the sea. Advantage was taken of this break in the construction of the line of railway from Madras to Beypore (Bépur); and it is through this gap that a great deal of the rain brought by the south-west monsoon in June and July escapes and falls in heavy showers on the eastern side of the Gháts over a wedge-shaped tract of country reaching to the coast. Cotton, tobacco, gram, millet, and the castor-oil plant are grown in abundance. All

the hills, except where cultivated, are thickly covered with forests, abounding in teak and other valuable timber, and frequented by great numbers of elephants, tigers, cheetahs, and other wild animals. The only important manufacture is that of cotton goods for exportation to the adjacent districts. Coimbatore (Kóyambuttúr) is well watered by the Káveri and its affluents, and shares in the rains brought by both monsoons.

*Towns.*—**Coimbatore** (Kóyambuttúr), the principal town, is situated at no great distance from the foot of the Nílgiris on the north side of the valley of Pálghát, and on the line of railway to Madras. **Dárápuram** is near the Amarávathi. **Bhavání**, at the confluence of the rivers Bhavání and Káveri, is a place of pilgrimage possessing two celebrated pagodas, one dedicated to Siva and the other to Vishnu.

[Coimbatore (Kóyambuttúr) was transferred to the British at the conclusion of the war ending in the death of Típu Sultan in 1799.]

69. **The Nílgiris.**—The Nílgiris (= Blue Mountains), from their central position, their great elevation, the salubrity of the climate, and the evenness of their seasons, have become the resort of Europeans from all parts of India, and of late years they have been made the seat of the Madras Government during the hot weather. The highest point, **Dodabettà** (= Great Hill), is 8600 feet above the sea. A large part of the surface of these hills is under cultivation, coffee, tea, and chinchona being extensively grown.

*Towns.*—**Ootacamund** (Utakalmand = One-stone Village), the chief station, is 7960 feet above the sea-level, and has an average temperature of 60° F. **Kótagiri**,

**Wellington**, and **Coonoor** (Kunúr), other stations on these hills, are respectively 6100, 5800, and 5700 feet above the sea.

[Besides the European residents, the Nílgiris are inhabited by several aboriginal tribes,—the Tódas, a handsome, tall, and muscular people, whose occupation is solely of a pastoral kind; the Kótas, who cultivate the soil and work as artificers; the Badugas, who, under tribute to the Tódas, are holders of the land; and the Kurumbas, a small and uncouth tribe held in dread by the Badugas as sorcerers. The Nílgiris produce coffee, tea, chinchona, and European fruits and vegetables.]

70. **Malabar** (Malaiyálam).—Malabar (Malaiyálam = Hill Country) is a narrow strip of land having the Western Gháts on the east side and the sea on the west. Harbours suitable for small vessels abound along the coast, which is low and sandy. The Gháts rise boldly from the plains to an elevation in the highest parts of 5000 or 6000 feet, and their sides are clothed with magnificent forests of teak, blackwood, and cedar, the timber of which is plunged into the torrents formed during the rains, and thus carried down to the coast. The passes are steep and difficult, but extremely picturesque. The soil is wonderfully fertile, and produces rice, cardamoms, coffee, and pepper in great abundance; the latter, indeed, forms so great a part of the wealth of the people, that it is sometimes called "the money of Malabar."

[In no part of India is the population composed of a greater variety of races than in Malabar (Malaiyálam); besides Europeans and foreign Asiatics settled on the hills and in the seaport towns, and the aboriginal races, there are Bráhmans, Náyars, Tíyars, Mápillas, Christians, and Jews. The Náyars are a very numerous and influential body, and were long the rulers of the country. They have very peculiar marriage customs. The Tíyars are chiefly cultivators. The Mápillas are Muhammadans, and carry on much of the coasting and Arabian trade. They are a keen, enterprising, and persevering race. Most of the Christians belong to the Syrian Church. The people of Malabar (Malaiyálam) are wholly engaged in trade and agriculture; there are no manufactures.]

[*Wynaad* (Vayanád), a division of this district, bounded on the north by Coorg (Kodagu), on the east by Mysore (Maisúr), on the south by the Nílgiris, and on the west by Malabar (Malaiyálam), is an elevated plateau, rising somewhat abruptly from the Malabar (Malaiyálam) side, but sloping more gently towards Mysore (Maisúr). It is an exceedingly picturesque tract, covered with verdure and flowers, and teeming with animal life. The soil is surprisingly fertile, and every little valley produces rice in abundance; plantains, jack-fruit, mangoes, and other common Indian fruits grow in profusion; cinnamon and nutmeg thrive well; and cardamoms without being cultivated spring up in great quantity in the jungle. The hill-sides are clothed with forests of magnificent growth; the iron tree, so called from the hardness of its wood, is of great girth, and rises to the height of sixty or seventy feet without a branch; and the white cedar grows to twelve feet in diameter. Coffee is produced in great quantity, and succeeds to perfection. Very recently gold has been found in the Wynaad.]

*Towns.*—**Calicut** (Kallikód) is the chief civil station in the district. It is a seaport town, and was formerly the capital of the possessions of the Zamorin. Vasco de Gama, the first European navigator who visited India, touched here in 1498. Here also in 1513 the Portuguese built a fort, and in 1616 the English established a factory. **Cannanore** (Kannúr) is a populous seaport town situated on a small bay having five or six fathoms of water. It has considerable trade in pepper, timber, and cocoanuts. Cannanore (Kannúr) was wrested from Típu by the British in 1791. **Tellicherry** (Talachéri) is beautifully situated on the seacoast. It has trade in pepper, cardamoms, and sandalwood. **Cochin** (Kochi), situated at the entrance of the magnificent lagoon extending north and south along the coast for about 200 miles, has considerable trade in the produce of the country. The approach to the harbour of Cochin (Kochi) is obstructed by a bar, but is practicable for vessels drawing fourteen or fifteen feet of water; inside the bar the water is thirty feet deep. Formerly Cochin (Kochi) was an exceedingly prosperous and wealthy place. **Pálghát** (Pálakkádu) is another town of importance; and **Manantoddy** (Mánantawádi) is the chief place in the Wynaad (Vayanád).

[Malabar (Malaiyálam) was, without provocation, invaded by Haidar in 1763. After plundering it to exhaustion, and carrying off 15,000 of the inhabitants, he evacuated it in 1768; but in 1773 he returned and again took possession. Típu, following the example of his father, also devastated the country, and treated the people with the most ferocious cruelty; but British troops drove his forces from it in 1792, and this part of India has ever since remained in peace under British rule. Malabar (Malaiyálam) with the whole of the coast region southwards is called *Kerala* by Hindu geographers.]

71. **South Canara** (Kannada). — South Canara (Kannada), like Malabar (Malaiyálam), is a narrow strip of hilly and very fertile country, watered by streams falling from the Gháts, and producing great quantities of rice, teak, sandalwood, pepper, ginger, and cardamoms. The coast is upwards of a hundred miles in length, with numerous inlets accessible only to fishing-boats or vessels of very small burthen.

*Towns.*—**Mangalore** (Mangalúr) is the chief town of the district. It was captured and recaptured several times during the wars between the British and Haidar and Típu, and fell finally into the hands of the former in 1798.

[South Canara (Kannada) was ceded to the British in 1799 by the partition treaty of Mysore (Maisúr), on the downfall of Típu.]

---

## COORG (KODAGU).

72. **Coorg** (Kodagu or *Kodumale* = Steep Mountains) is a very rugged region, the lowest part of which is more than 3000 feet above the sea. No part of the country is level; throughout it presents a constant succession of steep ridges of hills alternating with valleys or ravines, the whole being clothed with forests of the most stately growth. As indicated by the direction of the Káveri and its feeders, the country slopes towards the north-east and east. Cardamoms grow wild in great abundance; but coffee, which is extensively cultivated, forms the chief article of produce, and gives employment to great numbers of the people. Tea, chinchona, and cotton are also found to thrive well in various parts. The population is about 170,000 in all, but 30,000 only are Coorgs (Kodagas). These latter are superior in physical appearance to the people of the plains, and exceedingly brave. The country is governed by a British officer (styled "the Superintendent") appointed by the Government of India.

*Towns.*—**Mercara** (Madikéri), the chief town, is beautifully situated in the centre of the country, 3700 feet above the sea.

[In 1833, Viraréjendra Vodeyar, the reigning rája of Coorg (Kodagu), having violently oppressed his people and shamefully misgoverned his country, sought to put to death all the members of his family. His sister and her husband fled to Mysore (Maisúr), and claimed the protection of the British Resident. The rája demanded their return, but his demand was refused. Thereupon he made military preparations to enforce it. Negotiation was tried, and a native emissary was despatched into Coorg (Kodagu) by the British, but he was seized, detained, and insulted. Negotiation was therefore discontinued, and a British force marched into the country in April 1834, and captured Mercara (Madikéri); other strongholds also fell, and the rája unconditionally surrendered. He was deposed, and kept prisoner on a liberal pension, and the government of the country has ever since remained in the hands of the British.]

## TRAVANCORE (TIRUVÁNGKÚR).

73. **Travancore** (Tiruvángkúr) is a very well defined little country, physically divided into two regions—the low country and an irregular tableland. The former borders on the sea throughout, and nowhere exceeds fifty miles in breadth; south of Quilon (Kollam) the breadth is much less. The tableland separates the low country from the districts of Madura (Mathurai) and Tinnevelly (Tirunelvéli). On its western edge, this high land is about 2500 feet above the sea, but it rises gradually to the eastward, where it attains an altitude of from 4000 to 5000 feet. The climate of Travancore (Tiruvángkúr) is hot and moist, the mean temperature being 80° F., and the average rainfall seventy inches. Rain occurs almost throughout the year, and causes the country at all times to wear a verdant appearance. As is the case in Malabar

(Malaiyálam), rice, pepper, cardamoms, cocoanuts, ginger, betel, and areca-nut are largely produced, and all, except rice, are exported. Communication throughout the state is easy; there are many good roads, several crossing the mountains; and fourteen rivers, all navigable for small boats and rafts to the foot of the hills, cross the country from east to west, many of them falling into the backwater, thus affording easy access to the sea. The people are Nambúris or Bráhmans, Náyars, Syrian Christians, and Muhammadans; and number in all about one and a quarter million.

*Towns.* — **Trevandram** (Tiruvananthapuram = the Town of the Holy Eternal One), the capital, is a town of considerable size, and the residence of the Mahárája. **Aleppy** (Álappila) is the chief seaport town, from which great quantities of pepper, cocoanuts, cardamoms, and timber are exported. **Quilon** (Kollam) is also a considerable place on the sea-coast. The chronological era of the country dates from the foundation of this town, A.D. 640.

[Travancore (Tiruvángkúr) is governed by a Mahárája, and is one of the most progressive of the countries which remain under the rule of Indian princes. Police, hospitals, and schools have long been established. Missionaries are allowed to come and go without question; restrictions on commerce have been abolished, and the government is altogether so well regulated that several Englishmen of wealth have purchased land and settled in the country.]

## MYSORE (Maisúr).

74. **Mysore** * (Maisúr), 27,000 square miles in area, is enclosed on three sides by high ranges of mountains, having in many parts dense forests, extending from seven to ten miles from their base. The elevation of the country varies from 1000 to 6000 feet in height, but its general appearance is that of an undulating and elevated tableland, with isolated and detached ranges of hills of considerable elevation. The principal range of mountains is the Bába Budan range in the Nagar division. These hills are in the form of a horseshoe, with the opening towards the Western Gháts. A number of isolated hills or **droogs** (Sanskrit *durga* = an inaccessible place), usually rounded bare masses of granite about two miles in circumference, rising abruptly above the tableland, and fortified with successive lines of works rising tier above tier from the base to the summit, form a peculiar feature of the country. The chief of them are Nandidrug (*Nandi*, the name of Sivá's bull), Chittaldrug (*Chitra durg* = the Wonderful Fort), Severndrug (*Svarna durg* = the Golden Fort), and Astadrug. Mysore (Maisúr) is not a well-watered country, but it contains numerous streams, and tanks are common. The **Káveri** and its affluents the Tunga and Bhadra (Sanskrit *bhadra* = excellent), which unite

---

* Mysore (Maisúr) is a corruption of *Máhésh Asúra*, the name of a buffalo-headed monster, said to have been destroyed by the goddess Káli under the name of Chamundi, the tutelary deity of the Mahárája's family.

and form the **Tungabhadra**, the **Vedávati**, the **Pennár** (Pennaiyár), and the **Pálár**, are the principal streams. The soil produces rice, rági, millet, gram, wheat, sugar, and coffee. Mysore (Maisúr) is noted for its excellent draught oxen. Tigers, cheetahs, elephants, and other wild animals abound in the forests.

75. *People, Language, Industry.*—The population is estimated at 5,000,000. The Mysoreans are a robust race of people, taller and fairer than those of the Coromandel Coast. They are chiefly **Hindus** and **Muhammadans**, but the former greatly predominate. Their dress is fuller than that of the people of the low country, but their dwellings are similar, and their chief food is rági. The prevailing languages are **Kanarese** (Kannadam), **Telugu**, and **Hindustaní**. Cotton goods of inferior quality are made to some extent, but the principal manufactures are coarse black and white blankets, woollen carpets, and shawls.

76. *Government, Revenue.*—Mysore (Maisúr) is under the government of its own Mahárája. The revenue, which (as in other parts of India) is derived principally from the land, exceeds a crore of rupees.

77. *Divisions.*—Mysore (Maisúr) is divided into the three Divisions of NANDIDRUG, ASTAGRAM, and NAGAR. The first comprises the districts of **Bangalore** (Bengulúr), **Kólár**, and **Tumkúr** (Tumukúr); the second, **Mysore** (Maisúr) and **Hassan** (Hásanam); and the third, **Shimuga** (Sivamukhi), **Chitaldrug** (Chitradrug), and **Kadúr**.

78. *Towns.*—**Mysore** (Maisúr), the chief town of the

province, is the residence of the Mahárája. **Seringapatam** (Srírangapattanam = the Town of the Divine Vishnu), the capital of the province under Haidar and Típu, stands on the western extremity of an island in the Káveri. Its fort, which was so many times besieged, is now falling into decay; in the last siege of 1799 Típu was slain while defending it, and the place fell into the hands of the British. **Bangalore** (Bengulúr) is the most populous place in the territory, a large military station for British troops, and the terminus of an important branch of the Madras Railway. It stands on the high tableland of Mysore, and enjoys one of the finest climates in India. The mean temperature is about 76° F., and seldom exceeds 80° F. Numerous species of tropical and European flowers, fruits, and vegetables grow well in the neighbourhood. **Chitaldrug** (Chitradrug), the principal place of the district of the same name, is located at the base of a great rock surmounted by a strong fort. Other towns are **Kólár**, **Shimuga** (Sivamukhi), and **Tumkúr** (Tumukúr).

**[On the downfall of the house of Típu in 1799, Mysore (Maisúr), under its present limits, was erected into a separate state, and placed under the control of a rája of the ancient Hindu dynasty. But in 1832 the affairs of the country had fallen into such a state of disorder and confusion that rebellion ensued, and the British Government assumed the administration of the territory. Since then many evils have ceased to exist, and the country has steadily advanced in material prosperity. The present Mahárája was installed in March 1881.]**

## COCHIN (Kochi).

79. **Cochin** (Kochi).—Cochin (Kochi) is a small irregular-shaped hilly tract of country covering an area of 2000 square miles, between Malabar (Malaiyálam) and Travancore (Tiruvángkúr). Rice, pepper, cardamoms, cocoanuts, and coffee are extensively grown and exported; and the forests yield an abundant supply of teak and other valuable woods. Good roads are common in all parts of this state; and the backwater, which runs along the western side and sends its branches far into the interior, affords easy communication with the sea. The population, which is estimated at about 400,000, comprises Nambúris, Náyars, and Pulaiyars. These latter were, till very recently, a despised race and treated as slaves. Cochin (Kochi) is ruled by a rája who is responsible for all his actions to the British Government. The revenue amounts to about ten lakhs of rupees.

*Towns.*—**Trichoor** (Tiruchúr), situated near the backwater, is the next town in importance to Cochin (Kochi), which is British territory. It is the residence of the rája.

---

## THE FRENCH POSSESSIONS.

80. The French possessions in India are all, with the exception of Chandarnagar in Bengal, within the boundaries of the Madras Presidency. The principal territory is that of **Pondicherry** (Puduchéri), a tract a

little more than a hundred square miles in extent. The town of Pondicherry (Puduchéri), which is a clean, regularly built place, is the capital of French India, and the seat of its Government.

[The territory of Pondicherry (Puduchéri) was purchased by the French in 1672 from the King of Bijapur. The town itself has been taken and restored by the British four times, the last restoration being at the European Peace of 1815.]

81. The remaining possessions are **Karikal** (Kárakkal), a little town and territory in the District of Tanjore (Tanjávúr); **Yanaon**, in the Godávari District; **Mahe** (Mayyali), in Malabar (Malaiyálam); and **Chandarnagar**, on the Hugli, in Bengal.

---

## HYDERABAD (HAIDARÁBÁD).

82. **Hyderabad** (Haidarábád), called also "the Dominions of the Nizám of the Dakhan," from the title of the ruler, occupies nearly the centre of the peninsula of India. It is bounded on the north by the Bombay Presidency, Berár, and the Central Provinces; on the east and south by the Madras Presidency; and on the west by the Presidency of Bombay. The Wardha and Godávari rivers nearly mark its boundaries on the east, and the Krishna forms its southern limit. The area is estimated at 80,000 square miles; while the population falls little short of 10,000,000. The general surface of the country is regular and undulating, forming a slightly elevated tableland. Its climate is considered one of the most

pleasant and salubrious in India. During the greater part of the year the temperature is a delightful medium between the extremes of heat and cold experienced in the more northern parts of India. In the cold season, from the middle of November to the middle of February, the thermometer stands at 74° F.; in the hot months, from the latter period till the end of May, at 91° F.; and in the rains, from the early part of June till October, at 80° F. The average annual rainfall is thirty-two inches. In some parts the country is well cultivated, but brushwood covers the greater portion. Where irrigated, the soil produces cotton, wheat, and oil-seeds in great luxuriance; extensive topes of date and palmyra trees are found everywhere, but there is no extent of forest trees in any of the jungles. Several roads lead from Hyderabad (Haidarábád), the capital, to the coasts and other parts of the country. The Great Indian Peninsula Railway from Bombay, and the Madras Railway from Madras, meet together at Raichur, in the south-west corner of the state; and a third line, "The Nizám's State Railway," 121 miles in length, forms a branch line to Haidarábád, the capital. The revenue of the Nizám is stated to be about two crores of rupees.

83. *Towns.*—**Hyderabad** (Haidarábád), the capital, stands on the river Musi, a tributary of the Krishna. Its population is estimated at 200,000. **Secunderabad** (Sikandarábád), six miles north of Hyderabad (Haidarábád), is the largest British cantonment in India. In the vicinity is the extensive Husian Ságar, a tank several miles in circumference. **Bolárам**, in the neigh-

bourhood, is a cantonment for the Nizám's forces. **Warangal** was anciently the capital of Telingana, but is now a decayed place. **Nandér**, on the Godávari, is a place of pilgrimage for the Sikhs, whose warrior-guru, Govind, was here assassinated; a Sikh college is erected over the spot where the guru fell. **Jálna** (Jálnápur) is a British cantonment. **Aurangábád** (= the Place of the Throne), situated on the Dudhna, a tributary of the Godávari, is named after Aurangzéb, who was once Viceroy of the Dakhan. It contains the palace of Aurangzéb and the tomb of his favourite wife. The tomb resembles the Taj Mahál at Agra, but is built of coarser materials. **Daulatábád** (= Fortunate City), the Hindu *Deogarh*, is one of the most remarkable fortresses in India. At **Ellora**, a few miles to the north-west, are celebrated cave temples. **Assaye**, an insignificant village, is noted for the decisive victory there gained by 4500 British troops, under General Wellesley, over 50,000 Maráthas, the combined forces of Sindia and the Rája of Berár, 23d September 1803. **Bidar**, on a tributary of the Godávari, was formerly the capital of the province of Bidar. Ornamental hookah and betel bowls are manufactured here. **Golkonda**, six miles west of Hyderabad (Haidarábád), contains the tombs of the kings of Golkonda who ruled over this territory in the 16th and 17th centuries. Other large towns are **Gulbargah, Pangal**, and **Deverkonda**.

[On the dissolution of the Mughal empire, shortly after the death of Aurangzéb, many governors of provinces constituted themselves independent rulers. Amongst the number was Azof Jáh, Subahdár of the Dakhan, styled Nizám-ul-Mulk, or Regulator of the State. When he died in 1748, a fierce contest ensued for the succession. Názir Jang, the second son of the Nizám, seized the treasure, gained over the army, and prepared to en-counter Muzaffar Jang, his nephew, who claimed the throne as a bequest from his grandfather. Názir Jang's cause was espoused by the English, and that of Muzaffar Jang by the French. Názir, with the help of his allies, at first triumphed, but he yielded him-

View of the Royal Mausoleum, Golkonda.

self up to sensual pleasures, and quarrelled with his supporters for remonstrating with him, so that they withdrew shortly afterwards, and in conflict with the French he fell. Muzaffar Jang then became the undisputed ruler of the country, but he was assassinated almost immediately by the very native chiefs who helped him to the throne. The French then proclaimed Salábat Jang, younger brother of Názir Jang, and Ghází-ud-dín, the eldest son of the Nizám, opposed him. Ghází-ud-dín died suddenly, however, by poison it was thought, and Salábat Jang therefore remained in possession. For a short time matters seemed settled; but they were not so. Nizám Ali, another and younger brother, showed signs of uneasiness at the new arrangement, and to soothe him, Salábat Jang gave him a great deal of power. To strengthen his position, Salábat threw off the French, who were then scarcely able to maintain their own, and entered into an engagement with the English. Nizám Ali was not slow in showing his appreciation of the favours he had received from his brother; he first dethroned Salábat Jang, and then, two years afterwards, put him to death. This was in 1763. The wars with Haidar and Típu shortly followed, and on the downfall of their dynasty in 1799, the Nizám shared largely in the division of their territory, ceding his portion to the British for a large military force, maintained for his use at their expense. Nizám Ali died in 1803, and was succeeded by a feeble prince, who died in 1829. The affairs of the country continued to be badly administered under his successor, and a large debt to the British having accumulated, with little probability of its ever being discharged, in 1853 Berár was assigned to them for its payment and the maintenance of a military force.]

---

# THE HYDERABAD (HAIDARÁBÁD) ASSIGNED DISTRICTS.

84. *Physical Features, &c.*—These districts, otherwise known as **Berar**,* form the northernmost portion of the Hyderabad (Haidarábád) state. They have a total area of 17,000 square miles, and consist mainly of

---

* "Berar" is a corruption of *Vidarbha*, the name of a country (of which the present province formed a part) in the time of the "Mahábhárata." Damayanti, the wife of Nala, was the daughter of Bhíma, Rája of Vidarbha.

a broad valley running from east to west between the Sátpura range on the north and the Ajanta range in the south. This valley is watered by the Purna and a perfect network of streamlets descending into the main streams both from the hills in the north and the hills in the south. The soil is very fertile, producing cotton, millet, and other food-grains in abundance.

85. *People, Religion, Language.*—The people are about two and a quarter millions in number; are mostly **Hindus** in religion, and speak **Maráthi.**

86. *Government.*—Under the treaty with the Nizám by which these districts passed under British rule, the *sovereignty* of the province belongs to the Nizám of Hyderabad (Haidarábád), but the chief executive authority is vested in the British Resident at the court of the Nizám, who is immediately subordinate to the Government of India.

[The Resident, who is practically "Chief Commissioner" of the province, is assisted by a Revenue Commissioner, a Judicial Commissioner, and other responsible officers, and all the departments of the Government are managed on the same principles, laws, and rules as prevail in other provinces under British rule.]

87. *Revenue.*—The revenue amounts to about eight lakhs of rupees annually.

[The revenue is charged with the maintenance of the British forces known as "The Hyderabad (Haidarábád) Contingent," and the defrayal of all the expenses of the government of the province; all surplus revenue—at present about ten lakhs of rupees per annum—is paid over to the Hyderabad (Haidarábád) state.]

88. *Divisions.*—The province is divided into six districts, viz., **Amraoti, Akola, Ellichpur, Buldana, Wun,** and **Basim.**

*Towns.*—**Akola** is the chief town; **Amraoti** and **Ellichpur** are other towns of importance.

[When these districts came under British management in 1853, they were in a miserable state; the condition of the cultivators was most wretched, and violent crime was common. There were few roads, no bridges, and no police. Of late years, however, ruined irrigation works have been restored; tanks made; roads and railways formed; bridges constructed; hospitals, schools, and police established; and marked improvement generally has been effected.]

---

## THE CENTRAL PROVINCES.

89. *Position.*—These provinces stretch from Bundelkhand in the north to the Madras Presidency in the south; from the frontier of Bengal in the east to Málwa and the Dakhan in the west. Their area is 113,000 square miles.*

90. *Physical Features.*—These provinces are infinitely varied in local and topographical details, sometimes flat and fertile, but generally wild and rugged, abounding in hills, forest, and brushwood; sparsely populated, and scantily cultivated for the most part, but occasionally opening out into long and broad tracts covered with harvests and thickly inhabited; on the

* This includes the area of the feudatory state of Bastar (13,000 square miles), and also of fourteen other small protected states (16,000 square miles in all), under the political control of the Chief Commissioner of these provinces.

whole, poor and unproductive at present, but rich in natural resources, and capable of indefinite development in the future. The most prominent hill range is that of the **Sátpura Mountains**, south of the Narbada (Narmadá), and running parallel with it from east to west from its source on the tableland of Amarakantak to the westernmost borders of Nimár. The total length of the range is about three hundred miles, with an average breadth of sixty, making a mountainous region of about 18,000 square miles in extent. The other hill ranges are of less elevation. The chief plains are the open valleys of the Wardha and the Wainganga, both very fertile tracts, the former being the cotton field and the latter the rice field of these provinces, and together forming a plain 20,000 square miles in extent. The principal rivers are the **Narbada** (Narmadá), the **Mahánadi**, and the **Godávari**; but the first two, so far as their courses lie in these provinces, are quite unnavigable.

91. *Climate.*—There are three well-defined seasons—the hot, the rainy, and the cold. The first begins in April and lasts till the middle of June, when the second begins and lasts till October; the third continues from November to March.

[The temperature varies greatly at all times of the year; during the hot weather, the heat at noon averages 100° F., while the mornings and evenings are comparatively cold; and in the cold season the mean temperature is 40° F., but it sometimes suddenly rises to 80° F., and then as suddenly falls. Fogs and hail-storms occur at such times; ice also occurs in the cold weather.]

92. *Natural Productions.*—The mineral resources

are very great; **iron** is almost everywhere abundant, and the extensive jungles afford an unceasing supply of charcoal to smelt it. **Coal** of an excellent quality is worked in several localities near the Narbada (Narmadá), and at Chanda on the Wardha. **Marble, laterite,** and **limestone** abound; and **gold** is washed from some of the rivers. **Cotton, rice, wheat, Indian corn** and **millet, oil-seeds, opium, sugar-cane, safflower,** and **indigo** are grown in great quantities. Cotton of the finest quality is produced in the valley of the Wardha and Chhattisgarh; much is also raised in the valleys of the Narbada (Narmadá) and the Mahánadi. **Lac-dye** abounds in the forests. **Fibrous, medicinal,** and **edible plants** are found in great abundance, as well as trees yielding **timber, resins, gums,** and **dyes.** The Mahuá tree, which is spread over the face of the hilly districts, furnishes the people with food, oil, and an intoxicating drink. Oxen, buffaloes, sheep, and goats are the chief domestic animals; wild animals, innumerable birds, reptiles, and insects swarm in the forests; and fish of all sorts are caught in the tanks and streams.

93. *People, Religion.*—The population is over nine millions. Compared with other parts of India, this population is very scanty, the mean density being only seventy-nine to the square mile. **Hinduism** and **Muhammadanism** are the principal religions, but amongst the Gonds, or aboriginal tribes, the grossest ignorance and superstition prevail.

[Rude stones smeared with ochre, or upright posts with rough carvings at top, represent the Gond deities; and the malignant goddess Káli, under the name of Mahádevi (= the Great Goddess), is especially the object of their worship. Formerly human sacrifices were common amongst the Gonds, but owing to the vigorous administration of the British, the abominable rites have been, it is believed, suppressed.]

94. *Language.*—In Nágpur, **Maráthí** is used; in Sambalpur, **Hindí**; in the Godávari country, **Telugu**; and in the Gond districts, a great variety of dialects, each of which may be called a jargon of all three.

95. *Education.*—Public instruction is under the control of a Director. The schools amount altogether to about 2000, with 84,000 scholars.

[Most of the schools are as yet of the humblest kind. The cost of the higher establishments, of the grants-in-aid, and of the Departmental Establishment, is defrayed by the state; the cost of the village schools is defrayed from a cess amounting to one per cent. on the land revenue. The annual expenditure is rather more than five lakhs of rupees.]

96. *Industry.*—With the exception of weaving, and the making of brass and copper utensils, there are no arts or manufactures. **Agriculture** is almost the sole employment of the people, and even this is in many parts carried on the rudest possible way.

[In the forest districts ploughing is quite unknown. The inhabitants select a piece of ground on a moderate slope covered with trees, brushwood, and grass; the trees are cut down in November, the brushwood and grass are set fire to in May, and the charred ground is left covered with ashes. In the beginning of June, quantities of seed are placed at the upper end of the slope; the rains descending, wash the seed over and into the prepared ground, and there springs up a plentiful crop. About two millions of the people of these provinces depend on this method of cultivation.]

97. *Internal Communication.*—The means of communication are, as yet, few in proportion to the requirements of the country.

[Of late years the most strenuous efforts have been made in the construction of roads and bridges, in projecting canals, tramways, and railroads, and otherwise improving the means of communication. The main line of the *Great Indian Peninsular Railway* runs from Bhusáwal north-east through the Province to Jabalpur, and a branch from the same station runs eastwards through the Haidarábád Assigned Districts to Nágpur. Telegraphic lines have been constructed, and large sums of money have been spent in improving the navigation of the Godávari.]

98. *Commerce.*—The trade of these Provinces is carried on chiefly with Bombay, though that with the North-West Provinces, Central India, Bengal, Madras, and Berar is also considerable. The chief *imports* are cotton goods, hardware, and salt; the *exports* chiefly food grains, seeds, cotton, and leather.

99. *Government.*—Under the Government of India the administration is conducted by a **Chief Commissioner.**

100. *Revenue.*—The revenue is derived from taxes on land, salt, and sugar; stamps and excise. It is rapidly increasing, and amounts at present to about **a crore and twelve lakhs of rupees** a year.

101. *Condition of the People.*—Owing to the absence of education, to the want of means of communication, and to long years of misrule, the people of these provinces are, on the whole, **the most backward of any in India.**

[Many are in a perfectly barbarous state, dwelling in rude huts in the forests, and subsisting chiefly by the chase, while those in towns even are for the most part very uncouth. Crime, which was once fearfully prevalent, is still common; but under their present government these provinces have made and are making vast strides in moral and material improvement.]

102. *Divisions.*—These are:—(1) The TRANS-NARBADA districts, *Ságar* and *Damoh;* (2) the NARBADA VALLEY districts, *Jabalpur*, *Narsinghpur*, *Hoshangábád*, and *Nimár;* (3) the SÁTPURA HILL districts, *Mandla*, *Balaghát*, *Seoni*, *Chhindwara*, and *Betul;* (4) the NÁGPUR PLAIN districts, in the valleys of the Wardha and Wainganga, *Nágpur*, *Bhandara*, *Chanda*, and *Wardha;* (5) the CHHATISGARH districts, *Raipur* and *Bilaspur;* and (6) the SAMBALPUR district on the Mahánadi. The principal native feudatory states are *Bastar*, *Karond*, *Patna*, and *Bamra*.

103. *Towns.*—**Nágpur** is the seat of the Government of these provinces. Chintzes, coarse blankets, tent cloths, and copper and brass utensils are manufactured. Close to the city is the ridge of Sitabaldi. **Kamthí** is a large British Cantonment near Nágpur. **Jabalpur**, on the Narbada (Narmadá), is a thriving place. Coal of an excellent description, and fossil remains of the elephant and other gigantic quadrupeds, are found in the neighbourhood. **Ságar** is situated on a beautiful lake in a hilly district. **Burhanpur**, in Nimar, **Raipur**, **Chanda**, **Khandwa**, **Narsinghpur**, and **Hoshangábád** are other places of importance.

[The Ságar and Narbada (Narmadá) territory was ceded to the British by the Rája of Nágpur in 1818; the Nágpur country was annexed to the British Dominions in 1853, on the death, without heirs, of the Rája Rágóji. To these were added Sambalpur, and several petty districts forfeited by the acts of their rulers during the Mutiny of 1857, and the whole was constituted a separate government, under the name of the Central Provinces, in 1861.]

---

## THE LOWER PROVINCES OF BENGAL.*

104. The territory under the Government of the Lower Provinces of Bengal comprises the old provinces of Bengal, Bihár, and Orissa, Chutia Nágpur, and the tributary states thereto attached, the tributary states of Kuch Bihár and Hill Tippera, and the protected native state of Sikkim.

105. *Boundaries.*—These provinces are bounded on the north by Nepál, Sikkim, and Bhután; and on the east by Assam (Asam), the Lusháí Hills, the Arakan Hills, and the Náf River. The south is washed by the bay which derives its name from the province. At a point near Ganjám, on the sea-coast, the boundary line takes a westerly direction, and proceeding northward, verges in succession on the Madras Presidency, the Central Provinces, the independent state of Rewá, and the North-West Provinces. The area of the whole is estimated at about 188,000 square miles.

---

* Bengal is the ancient *Bangála*, or *Banga Désa.* It was also called *Gaur.*

106. *Physical Features.*—The physical features of the Bengal Presidency are for the most part clearly defined, and exercise an important influence over the nationality, character, and occupations of its inhabitants; its rivers are of remarkable size, its hills mainly in significant positions, and the plains of vast expanse and considerable fertility. Broadly speaking, the chief characteristics of these Provinces are the valley of the Ganges proceeding from the north-west, and the valley of the Brahmaputra from the north-east, meeting at the base of a V, and spreading a network of minor streams over the land which forms their delta. On the west rise the high lands of Chutia Nágpur; and on the east the Chittagong (Chattagrám) and other hills. To the north, between the courses of the great rivers and the Himálayas, and enclosed by the Tarái at the base of the latter, is a large tract of open country, having little variety of aspect, and a number of minor rivers, all of which flow into one or other of the two great streams.

107. *Climate.*—The seasons in Bengal are very well defined, being divided into the hot, the rainy, and the cold. The hot weather lasts from the middle of March to the middle of June, the rains from thence to October, and the remainder of the twelve months is called the cold weather.

[During the hot season the thermometer frequently rises to 100° or 110° F. When the heat is greatest, the earth is parched up and rent, the sun shines down with great intensity, not a cloud is to be seen, not a particle of air is in motion; flowers and leaves droop, animals seek the shade, and mere existence seems fatiguing.

At night the excessive heat decreases, but next morning it returns with the same intensity, and a shower of rain is felt as the greatest blessing. In the cold season, north winds prevail, and shallow pots of water placed in the open air all night are found in the morning covered with thin sheets of ice. In the rains, the valleys of the Ganges and Brahmaputra are covered with water, and the country appears as an inland sea, dotted with numbers of small inhabited islands. Owing to the profuse exhalations from the stagnant waters and decaying vegetation, this season is the most unhealthy in Bengal. The average annual rainfall is about 70 inches.]

108. *Natural Productions.*—MINERAL PRODUCTS.—Almost the whole of these provinces has been formed by soil brought down by the rivers from the distant mountains, and black mould and sand appear for hundreds of miles; nothing so coarse as gravel occurs, so that it is said "it is impossible to find a stone sufficiently large to throw at a dog." **Nitre** is largely mingled with the soil. **Coal** is worked at Rániganj on the Dámudar. VEGETABLE PRODUCTS.—The luxuriance of vegetation in Bengal is not surpassed, perhaps, in any part of the world. The country may be considered as one enormous garden. **Rice** is most generally cultivated, but wheat and many other food-grains are grown. Poppies, from which **opium** is manufactured, **indigo, jute, cotton, sugar-cane, mulberry,** and **tobacco**, form the principal commercial crops. **Oil-seeds, ginger, chillies,** and other esculent vegetables; mangoes, oranges, limes, plantains, and other **fruits** are produced in profusion, and everywhere beautiful flowers and gorgeous flowering trees and shrubs grow without care. The Súndarbans (Sanskrit, *súndari vana* = Forest of Súndri Trees), and the forest at the base of the

Himálayas, produce great quantities of useful **timber.** ANIMALS.—These are extraordinarily numerous and varied. **Tigers** of the greatest size and ferocity infest all the jungles; **elephants** are common in the forest belt at the base of the Himálayas; **hyenas, wolves, bears, jackals, deer,** and **monkeys** are also common. **Reptiles** of all kinds are exceedingly numerous, and all the rivers are infested with **crocodiles.** Crows, kites, sparrows, and other **birds** are met with throughout; waterfowl is abundant; and amongst large birds is "the adjutant," a gigantic crane which frequents towns in considerable numbers to carry off the garbage. **Fish** is very plentiful, and insects of all kinds are numerous. The domestic animals are, however, very inferior.

109. *People, Religion.*—The population comprises several distinct nationalities, each mainly resident in one of the different provinces named. Bengal is inhabited throughout by Bengalí-speaking people to the number of 34,000,000. The people of Bihár are Hindustánís identical with the people of the North-West Provinces, and number about 20,000,000. The Uriya-speaking population occupy Orissa, and are in number about 4,000,000. The bulk of the people are **Hindus,** but there are no less than 20,000,000 **Muhammadans.**

[The Santáls, a primitive race inhabiting the wild jungly tracts in Bhágalpur and Bírbhúm, are about 1,250,000 in number.]

110. *Language.*—The languages are **Bengálí** in Ben-

gal, **Hindustání** and **Hindí** in Bihár, and **Uriya** in Orissa. English is everywhere making rapid progress.

111. *Education.*—Public instruction is under a Director. Education is most flourishing in and around Calcutta, the centre of commerce and of Government; but it is everywhere rapidly extending.

[The annual cost of public instruction amounts to more than thirty lakhs of rupees.]

112. *Industry.*—In Bengal, as elsewhere in India, **agriculture** forms the chief occupation of the people, but the **manufactures** are many and important; silk and cotton goods of various descriptions, chintzes, blankets, and sailcloth are amongst the chief; and large quantities of a coarse cloth are made from jute. The preparation of opium, indigo, silk, and sugar, and the making of jewellery, saddlery, perfumes, ornamental ware, and pottery, give employment to thousands of people.

113. *Internal Communication.*—No country of the same extent has so many navigable rivers. The Ganges with its tributaries forms the great highway for most of the commerce of the country; the busy scene which it constantly exhibits, and the number and variety of its boats, are not to be equalled, perhaps, on any other stream in the world.

[The navigation of the Ganges is, however, always a work of difficulty. Against the stream, boats can only proceed by sailing or tracking. If the wind be contrary, the latter operation must be performed; most of the crew then go on shore, and, with ropes tied to the mast, drag the vessel along from morning to night, the

Boats and Boatmen, River Ganges.

men wading through mud and water. Sometimes the boat gets aground, and remains so for hours. Coming down stream, the same thing constantly happens, and occasionally the boat capsizes. Before the introduction of steamers and railways, travellers of all ranks were obliged to proceed in this way. The steamers on the Ganges are specially constructed for its navigation, being made to draw little more than three feet of water. When the river is full the downward voyage is made with great rapidity.]

[ROADS.—Roads in Bengal are not so common as in other parts of India; materials for making them do not exist, and the cost of construction therefore is great. Calcutta is connected with neighbouring towns by causeways constructed of brick, and raised far above the level of the surrounding country, to be beyond the reach of the floods. Railroads are dealt with in the same way. When the country is covered with water, the villages with their huts, granaries, trees, and cattle, seem to be rising out of a lake; the ordinary tracks are then several feet below the surface, and boats are the only means of transit which the villagers can use.]

[RAILWAYS.—The "Eastern Bengal Railway" runs from Calcutta to Goalando, a distance of 156 miles; and a short line of 28 miles runs southwards to Port Canning. The "Northern Bengal State Railway" runs from Poradaha on the Eastern Bengal Railway, to Siliguri at the foot of the Dárjiling Hills. The "East India Railway," the longest line in India, is open to Dehli, a distance of 1018 miles. It passes through Chandarnagar to Bardwán, whence a branch proceeds to the coal-mines of Rániganj, while the main line runs northward to the Ganges, nearly keeping the river's course, into the North-West Provinces, and placing all the large towns in communication with Calcutta.]

114. *Commerce.*—IMPORTS.—The chief imports are **cotton, woollen,** and **hardware goods, machinery, manufactured metals, glass, wines, spirits, beer,** and **salt.** EXPORTS.—These are principally **opium, rice, wheat,** and other food-grains, **jute, cotton, indigo,** and other dyes, **raw silk** and **silk goods, oil-seeds, sugar, tea, saltpetre,** and **hides.** The countries chiefly traded with are England, France, and China. A great deal of commerce is also carried on with Bombay. PORTS.

—Calcutta, Chittagong (Chattagrám), and Bálasore (Báléshwar).

115. *Government.*—These provinces are subject to the direct control of a **Lieutenant-Governor.**

116. *Revenue.*—The revenue, which is derived from land, opium, customs, stamps, excise, and sundry minor sources, amounts to about **seventeen crores of rupees.**

[The land tax yields about four crores of rupees annually: opium somewhat more, and customs three crores.]

117. *Condition of the People.*—In Bengal the ordinary habitations of the people are built of mud, or of brushwood plastered over with mud: the frames consist of bamboos tied together; and the roofs are constructed of two sloping sides which meet at the top in a ridge. There is no window or aperture except the doorway, and smoke escapes how it can. Each hut is concealed by thick vegetation to keep off the sun, and each has a patch of ground in which vegetables are reared; these, rice, and occasionally fish, form the ordinary food. The better classes wear a long cotton robe, with a white scarf tied round the waist. The dress of the females is very elegant. The hands and feet are usually adorned with jewelled ornaments, and even the women of the working classes have their armlets and anklets of brass or silver.

[Many of the Zamindars, or great landed proprietors, are exceedingly rich, but the peasantry are very poor. These last are, however, content with little, and where they reside at a distance from towns are as innocent, temperate, and moral as the people of any country in the world. In Bihár the houses are roofed with tiles, and built in compact villages in open tracts.]

118. *Divisions.*—The provinces of Bengal, Bihár (*Vihár*, a Buddhist monastery), and Orissa are divided into a number of small but populous districts, grouped into the following Divisions, technically called "Commissionerships:"—**Patna, Bhágalpur, Rájsháhi** and **Kuch Bihár, Presidency, Bardwán, Orissa, Dacca** (Dháká), and **Chittagong** (Chattagrám). To these must be added the Commissionership of **Chutia Nágpur.**

119. CALCUTTA, situated in N. lat. 22° 33′, E. long. 88° 19′, the capital of Bengal, and the seat of the

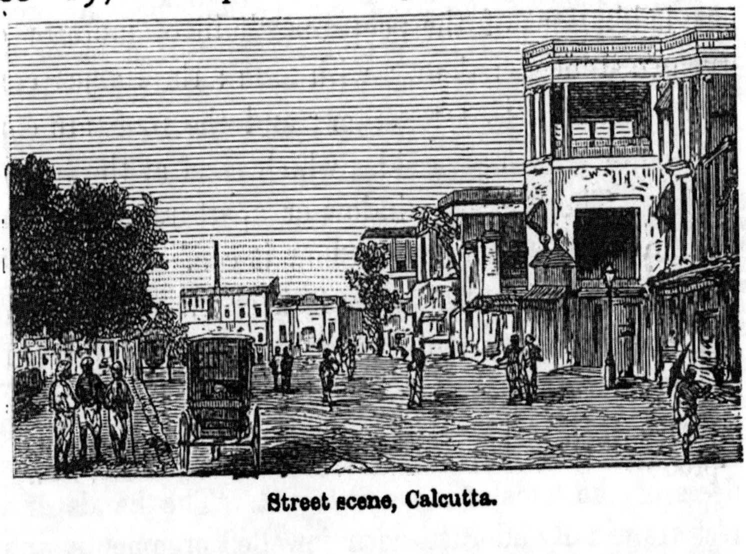

Street scene, Calcutta.

Government of India, and of the Government of the Lower Provinces, stands on the left bank of the Hugli, about a hundred miles from the sea.

[It is divided into two distinct portions, the northern or native portion having narrow, crowded streets, and mean houses; while the southern or European portion has spacious streets, and edifices so large and handsome as to gain for Calcutta the name of "The City of Palaces." The suburbs are extensive, and laid out in out in parks, in which stand the houses of the wealthier classes. Fort William, constructed by Lord Clive after the battle of Plassey, is one of the largest fortresses in India, and capable of holding 15,000 men. Opposite the city, the river, here about half a mile wide, presents a most animating scene ; scores of ships, of all sizes and from all countries, are anchored in the stream, and hundreds of native craft of all descriptions ply in every direction. Railways connect Calcutta with other parts of Bengal, telegraphs with the principal parts of India, steamers with other parts of Asia, and ships with all other parts of the world. The population of the city and its suburbs is estimated at 700,000.]

120. **Patna Division.**—This division comprises the districts of **Patna, Shahábád, Gayá, Darbhangah, Muzaffarpur, Sáran,** and **Champaran.** Nearly the whole of Patna is perfectly flat and subject to inundation. The soil is very fertile, highly cultivated, and well watered. Rice, wheat, and barley are extensively grown, and much opium is produced of the finest quality. Shahábád presents considerable variety of physical feature, from the flat alluvial plains, over which the annual inundation spreads its coating of silt, to the rocky hills in the south-west. Besides the usual grains, great quantities of cotton, opium, indigo, and betel are grown. Gayá closely resembles Shahábád in the variety of its physical features. Rice of excellent quality is largely produced. The numerous torrents which traverse and inundate this district render it very difficult to form good roads and keep them in repair; it is therefore almost destitute of them. Much of the districts of Darbhangah and Muzaffarpur is subject to inundation, but the greater part is formed of undulating ground. The whole is of great fertility, well watered, and beautifully wooded. There are many good roads, and great facilities for water-carriage and irrigation

Though these two districts are together little more than 6000 square miles in extent, they have the immense population of 4,400,000. Sáran and Champaran closely resemble the above in physical aspect and fertility. They are traversed by the Gandak (Gandakí), the Gogra (Sanskrit *Gharghara*=making a gurgling sound), and the Ganges.

*Towns.*—**Patna** (the ancient *Pátaliputra*), the principal place in the division, has much trade. Population 159,000. It was here that, in 1763, two hundred Englishman were barbarously put to death by Sumru, the agent of Mír Kásim. **Gayá**, the civil station of the district, abounds in shrines, which are visited every year by thousands of pilgrims. It was here that Gautama or Buddha (the Enlightened), the founder of Buddhism, first taught his system of religion. Other important towns are **Arrah, Darbhangah, Muzaffarpur,** and **Chupra.**

121. **Bhágalpur Division.**—This division comprises the districts of **Monghir** (Múngír), **Bhágalpur, Purnía, Máldah,** and the **Santál Parganas.** Like Patna, it includes a great stretch of country from the foot of the Himálayas to the Ganges, and thence to the hilly tracts on the south. The district of Monghir (Múngír) is completely cut in two by the Ganges. All to the north of that river is subject to annual inundation, but the southern portion is more elevated and beyond its influence. The staple crops are rice and wheat, but opium, oil-seeds, indigo, sugar, and tobacco are largely produced. Bhágalpur (the ancient *Anga*) is traversed by the Ganges. The southern portion is hilly, in some

parts cultivated, in others covered with an almost impenetrable jungle infested with wild beasts. Numerous

Rock sculpture, Mount Mandara,

torrents rush down from the hills during the rains and flow northward into the Ganges. The chief crops are rice, wheat, cotton, sugar-cane, and indigo. Of the last, large quantities are exported.* Purniá† and Máldah lie on the north side of the Ganges.

*Towns.*—**Monghir** (Múngír) is on the Ganges. **Bhágalpur** is the principal town in its district. A little to the north-west of it are two singular round towers, supposed to be of Jain origin, and meant for the accommodation of the numerous worshippers of this sect who annually visit them.

[SANTÁL PARGANAS.—These, for the most part jungly tracts inhabited by Santáls, are partly in Bhágalpur and partly in Bírbhúm.]

122. **Rájsháhi** and **Kuch Bihár Division.**—This division embraces the districts of **Dinájpur, Rájsháhi, Rangpur, Bogra** (Bagurá), **Pábná, Dárjiling, Jalpaiguri**, and **Kuch Bihár**. The Rájsháhi portion of this division is little more than one vast flat, subject to annual inundation. Numbers of navigable streams give it great facilities for water-carriage, and innumerable jhíls, or small stagnant sheets of water formed in the deserted channels of streams or by the overflowing of springs, are found spread over its surface. The numbers and position of these constantly vary, old ones are silted up and new ones are formed. The Kuch Bihár portion comprises the districts of Dárjiling,

---

* A few miles south of Bhágalpur is the famous Mount Mandara, a granite hill about 700 feet high, abounding in ruined temples and rock sculptures. It is a great place of pilgrimage, as, according to the Puránas, it was with this hill that "the gods churned the milky sea."

† Purniá formed part of the ancient *Mithila*, and, according to Hindu tradition, was governed by a rája whose daughter was Sítá, the wife of Ráma. Her abduction by the giant Rávana of Lanka brought about the war related in the "Rámáyana."

Jalpaiguri, and Kuch Bihár. The Dárjiling Hills are well adapted for growing coffee and tea, the latter of which is there extensively cultivated.

*Towns.*—**Dinájpur**, **Rangpur**, **Pábná**, **Dárjiling**, **Jalpaiguri**, and **Kuch Bihár** are the principal civil stations in their respective districts. Dárjiling is used as a sanitarium. **Bogra** (Bagurá), the chief place in its district, stands on the banks of the Korotya, a river which Hindus used to consider the limit of civilisation, and beyond which it was not lawful for a Bráhman to pass. **Rámpur Bauleah** is the chief town in Rájsháhi.

123. **Presidency Division.**—The Presidency division constitutes the central portion of the great Gangetic delta, and comprises the districts of **Murshidábád**, **Nuddea** (Nadíyá, *Nava dwipa* = New Island), **Jessore** (Jasohar), **Calcutta**, and the Twenty-four Parganas. This division does not differ in any important respect from the flat districts already noticed, but the southern portion, known as the Súndarbans, consists of low-lying marshy tracts and impenetrable forests, broken up by innumerable tidal estuaries. Many of these latter are navigable for large vessels, but the mudbanks by which they are lined are constantly shifting, and except the Hugli there is hardly another channel out of the many which a vessel may safely trust to. The thickets are infested with tigers and other wild animals, and swarms of alligators abound in the water.

*Towns.*—**Berhampore** (Barahampuram), the chief town in the district of Murshidábád, stands on the left bank of the Bhágírathí, a great offset of the

Ganges. **Murshidábád**, also on the Bhágírathí, was formerly the capital of Bengal; it is still a place of considerable trade. **Krishnágarh** and **Jessore** (Jasohar) are the chief towns in Nuddea (Nadíyá) and Jessore (Jasohar) respectively.

124. **Bardwán Division.**—This division includes the districts of **Bírbhúm** (*Víra bhúmi* = the Land of Heroes), **Bardwán, Bánkurá,** Midnapur (Medinípur), and **Hugli** with **Howrah.** The Hugli forms its eastern boundary; and on the south it reaches the sea. Much of it is flat and liable to severe floods, but westward the country rises and breaks into undulations. Like the rest of Bengal, it abounds in jhíls, streams, and navigable rivers. The Bhágírathí and the Hugli are navigable at all times; and in the rainy season the Dámudar (*Dámodara,* one of the names of Vishnu) and other streams serve as channels for the conveyance of the coal of Bánkurá and the iron of Bírbhúm to Calcutta. These districts are amongst the most fruitful in India; rice, sugar, oil-seeds, indigo, and all the other Bengal products being yielded in profusion. Much salt is made in places near the sea, and the fisheries are very important. A range of low sandhills runs along the shore, and where it fails, the sea is kept from flooding the adjoining country by artificial embankments.

*Towns.*—**Bardwán** (*Vardhamána-pura* = Thriving Town), the principal place in its district, is situated on the Dámudar. **Bánkurá** has manufactories for the preparation of indigo and lac-dye. **Raníganj,** on the Dámudar, has coal-mines in the vicinity; the coal is found in veins of about eight feet in thickness, and there are more than forty collieries.

125. **Orissa Division.**—This division includes the districts of **Bálasore** (Báléshwar), **Cuttack** (Katak), **Puri**, and the **Orissa Tributary Máháls.** It lies on the north-west of the Bay of Bengal, and has a sea-board stretching from the north of the Subanrékha (*Suvarna rékha,* streak of gold) to the northern extremity of Ganjám. Much of the coast is rendered dangerous by sandbanks; the country itself along the shore is for the most part flat and sandy, but gradually rises towards the west into hills exceedingly rugged and jungly. The delta of the Mahánadi is a low swampy tract resembling the Súndarbans, overgrown with coarse grass and brushwood and intersected by numerous tidal creeks. The whole country is well watered by the Mahánadi, Bráhmani, Baitarani, and many smaller streams. In the hot season these latter are dried up, but in the rainy season they swell to large rivers. Rice, pulse, sugar-cane, oil-seeds,

indigo, and cotton are grown in the low lands; wheat and tobacco in the more elevated portions; and timber, chiefly teak and sál, drugs, dyes, resins, wax, honey, and wild fruits on the hills. The timber is floated down to the coast when the rivers are swollen. Both the plains and the hills are infested with great numbers of wild beasts and vast numbers of venomous reptiles. Iron is obtained in the forest tract; and the best and whitest salt made in India is manufactured along the coast and on the banks of the Chilká lake. The inhabitants are chiefly Uriyas and speak the Uriya language.

[*The Orissa Máháls.*—The native states called the *Orissa Máháls* lie to the westward; they cover an area of 16,000 square miles, and consist for the most part of barren rocky wastes and dense jungles inhabited by a semi-barbarous people.]

*Towns.*—**Cuttack** (Katak), the principal town, is situated on a small peninsula formed by the Mahánadi. **Bálasore** (Báléshwar), the seaport of the country, is situated on a river navigable for vessels of a hundred tons. Considerable quantities of rice are exported. **Purí** or **Jagannáth**, a small town on the coast, contains the celebrated temple of Jagannáth, held in great estimation both by Buddhists and Bráhmanical Hindus.

[Orissa is a corruption of *Odhradésa*, or *Ordésa*, the land of the *Ohhra* or *Or* tribes. The early history of the country is lost in absurd legendary tales, but authentic accounts begin about A.D. 470. A thousand years later, Rája Anang Bhím Deo, who erected the great temple of Jagannáth, ruled over the land. In 1558 the Government was wrested from its native rájas by the Afgháns, and from them again in 1578 by one of the generals of Akbar, who made it a part of his dominions. In 1743, during the weak reign of Aliverdi Khán of Bengal, the Maráthas invaded Orissa, and subjected it to the most frightful exactions and misrule and violence. In 1803 the British took it from the Rája of Berár, and added it to their possessions.]

126. **Dacca** (Dháká) **Division.**—This division includes the districts of **Bákharganj**, **Dacca** (Dháká), **Faridpur**, **Maimansinh**, **Tippera**, and **Hill Tippera.** Bákharganj, the most southern portion, forms part of the Súndarbans, and abounds in tidal creeks, jhíls, and jungles; but it is higher, less swampy, and more thickly inhabited than the districts farther to the west, and where cultivated produces two crops of rice a year. Inundations occur, and occasionally they do extensive damage. The whole of the country, from the bay to the northern parts of Maimansinh, is an uninterrupted flat traversed by many rivers and watercourses, and abounding in shallow lakes left by the annual floods.

*Towns.*—**Barisál**, the principal town of Bákharganj, is situated on a branch of the Ganges. **Dacca** (Dháká), on one of the branches of the Ganges, is a very ancient city, supposed to have been founded 1500 years ago. It was for some time the seat of the Muhammadan Government of Bengal, and till lately was celebrated for its muslins, which were of such singular fineness and beauty that they were called "webs of woven wind." Dháká still has some manufactures, but its muslins have been superseded by the cheaper fabrics of England. Near Dháká is the populous town of **Náráyanganj. Faridpur** and **Maimansinh** are the chief towns in their respective districts. Commillah is the chief town in Tippera.

127. **Chittagong** (Chattagrám) **Division.**—This division includes the districts of **Noakhali**, **Chittagong** (Chattagrám), and the **Chittagong** (Chattagrám) **Hill Tracts**, and is naturally divided into two distinct parts.

Noakhali has all the physical characteristics of Bengal; while Chittagong (Chattagrám) lies along the east coast of the bay, and stretches far into the hilly region beyond. In this latter district the fertile portions are chiefly on the coast and in the valleys between the hills, where large crops of rice, cotton, sugar-cane, tobacco, indigo, ginger, and betel, are raised with very little labour. Much tea is grown on the hills and exported. In the hilly tracts the rudest system of agriculture is practised by the natives; the side of a mountain is cleared of wood by fire; at the commencement of the rains small holes are chopped in the ground, into which are dropped paddy, gourds, pepper, and cotton seed; and the first up is first reaped. Timber of the most useful kinds abounds in the forests. Elephants are numerous, and are caught and trained to labour. Salt is made in considerable quantity all along the coast.

*Towns.*—**Chittagong** (Chattagrám), the chief town in the division, is situated near the mouth of the river of the same name, here a wide estuary navigable for large vessels. Rice and timber are largely exported.

**[Chittagong (Chattagrám) was wrested from the Arakanese by Aurangzéb, and in 1760 it was ceded to the British by the Nawáb of Bengal.]**

128. **Chutia Nágpur.**—Chutia Nagpore comprises the districts of **Lohardaga, Hazaribágh, Mánbhúm, Singbhúm,** and a number of smaller states known as the Tributary Máháls. It is hilly almost throughout, scantily populated, and has by far the greater part of its surface covered with jungle and forests. Much of

it is of great fertility, but being far from the sea, and a wild uncultivated region, crossed by numerous rivers and destitute of roads, its natural wealth cannot well be turned to account. Cotton is grown in some of the districts and carried on pack-bullocks to the marts of Bengal: immense quantities of teak, ebony, and other kinds of valuable timber, drugs, dyes, and gums are produced in the forests; and besides copper, which is worked near Chaibásá, in Singhbúm, coal, iron, and other minerals are known to exist.

*Towns.*—The chief towns are **Ranchi, Hazaribágh, Chaibása**, and **Purulia.**

---

129. **Hill Tippera.**—Hill Tippera is a very mountainous tract, almost covered with dense bamboo jungles, infested with elephants, deer, wild hogs, monkeys, snakes, and birds of all kinds. The government of the country is in the hands of a number of chiefs, who pay annual tribute, in ivory, goats, and cotton fabrics, to the rája; the chiefs levy tribute on their subjects at will.

[The people are known as Tipperas, and are a black undersized race, as wild as the country they inhabit. Both sexes wear nothing but a narrow cloth round the loins, and the men go constantly armed with bows, arrows, spears, and other rude weapons. They live together in societies of from twenty to a hundred, in widely scattered hamlets of huts constructed on platforms elevated from about four to seven feet above the ground; the roofs of the huts are thatched with grass, and the walls are of bamboo slips plaited. Under the platform, pigs, fowls, and other domestic animals are kept. Owing to their uncleanly habits and to the gross food which they take, four-fifths of the population are victims of some disgusting disease. They cultivate paddy, cotton, Indian-corn,

**and indigo, and pumpkins and yams are extensively grown. The mode of cultivation is peculiar. In March the bamboo jungle is felled, dried, and burned; after the first shower of rain, men, women, and children, by means of a rude hatchet, chop the ground into little oblique holes about three inches deep; into these holes the seed is dropped, and as each one drops in whatever seed he or she pleases, very commonly paddy, cotton, indigo, and maize may all be seen coming up together. The ground is so fertile that ploughing and manuring are not needed.]**

130. **Sikkim** is a small mountainous tract lying between Bhután on the east and Nepál on the west. In physical features it resembles the neighbouring parts of Bhután, and its productions are similar. It is governed by a rája.

---

## ASSAM (ASAM).

131. *Boundaries, Dimensions.*—The province of Assam (Asam) lies on the north-east border of Bengal. It is bounded on the *north* by Bhután, and ranges of hills inhabited by wild tribes; on the *north-east* by the Mishmi Hills, which sweep round the head of the Brahmaputra valley; on the *east* by the frontier mountain ranges of Burma (Barma) and Manipur; on the *south* by the Lushai Hills and Hill Tippera; and on the *west* by Bengal and Kuch Bihár. Its area is estimated at 42,000 square miles.

132. *Physical Features.* The province consists of (1.) the valley of the Brahmaputra; (2.) the valley of the Surmá or Barák River; and (3.) an intervening moun-

tain tract known as the Gáro, Khási and Jaintiá, and Nága Hills. The valley of the Brahmaputra, or Assam proper, is an alluvial plain about 450 miles long and

Scene on the River Surmá.

50 wide, shut in by mountains on the north, east, and south, with the Brahmaputra flowing through its entire length from east to west, and crossed by count-

less streams falling into the main river from the bordering mountains on either side. The uniform level of the valley is here and there interrupted by clumps of green conical hills from two to seven hundred feet high. During the rainy season the rivers overflow, and much of the country is then under water. The valley of the Surmá, so called because it is intersected by the Surmá or Barák River, comprises the two districts of Sylhet and Cachar (Kachhár). The western portion (Sylhet) closely resembles Assam (Asam) proper in physical appearance, but the eastern portion (Kachhár) is more diversified. Low offshoots from the hills bounding the valley both north and south extend almost to the river's edge on either side, with rich alluvial valleys between. Much of the low lands in these valleys is under water during the rains. The Central Hill Tract, formed by the Nága, Khási and Jaintiá, and Gáro Hills, is a continuation of the mountain chain which sweeps round the head of the Brahmaputra valley. The major portion of the whole of this tract is covered with dense jungle.

133. *Climate.*—The climate of Assam (Asam) is humid, the cloudless skies characteristic of India generally being but little seen in this province. The rains, which in most parts are very heavy,* commence in March and last till October; and the winds blow mainly from the north-east or east, this part of India being beyond the influence of the monsoons. The

* Cherápungí, in the Khási and Jaintiá Hills, has the greatest known annual rainfall in the world.

## BRITISH BURMA (BARMA).

**[The three maritime provinces of India beyond the Ganges, comprising the ancient kingdoms of Pegu and Arakan, with the long line of sea-coast called Tenasserim, were united under one local administration in January 1862, and called British Burma (Barma). The last two of these were acquired by treaty after the war of 1825–26, and Pegu was occupied and retained consequent on the war of 1852.]**

144. *Boundaries.* — British Burma (Barma) is bounded on the west by the Bay of Bengal; Arakan is bounded on the north by Chittagong (Chattagrám) and some independent states, and on the east by the Yoma Mountains; Pegu is separated from upper Burma (Barma) on the north by a line corresponding to the 19° 30′ parallel of north latitude, and is bounded on the east by the Salwen; Tenasserim is bounded on the east by a long line of mountains separating it from Ṣiam, and varying in height from 3000 to 5000 feet above the sea. The area of the whole province is estimated at 88,000 square miles.

145. *Physical Features.*—Arakan is separated from Pegu and Upper Burma (Barma) on the east by a range of mountains which attains at its greatest elevation a height of 7000 feet. The range runs nearly parallel with the line of sea-coast and gradually lowers towards the south. The northern portion of the country has a large extent of alluvial soil. In the lower course of the river Kuladan and its numerous affluents, the breadth of the land from the shore to the watershed mountains is from eighty to ninety miles. A large

portion of the country is hilly, covered with forest and difficult of access. The watershed range separating Arakan from Pegu extends southerly, and between that range and the seashore, for a length of nearly two hundred miles, as far as a point near Cape Negrais, the country is a mere narrow strip of land. Pegu and Martaban lie in the valleys of the Iráwadi and Sittang rivers. These valleys, bounded east and west by mountain ranges, are narrow in their upper portions, but expand at the delta of the Iráwadi into a magnificent alluvial region, penetrated by a vast number of tidal creeks, and extending over 10,000 square miles. Northern Tenasserim is bounded on the west by the last hundred miles of the course of the great river Salwen. This river, without a delta, and with passless rapids one hundred miles from its mouth, which bar the ascent of boats to Central Asia, is joined at Moulmein (Múlmaín) by two streams which facilitate communication with the interior. The breadth of land from the seashore at the mouth of the Salwen to the central mountain range which divides the province to the eastward from Siam, is about eighty miles; but this space is gradually narrowed to forty miles at its southern extremity. There is no river of importance south of the mouth of the Salwen. The interior of the country is a wilderness of hills, thickly wooded, with long narrow valleys running in a general north and south direction.

RIVERS.—The **Kuladan** rises in the mountains to the east of Arakan; the upper part of its course lies through a desolate and mountainous region, but in the lower part it traverses a plain dotted with villages and covered with luxuriant crops of grain. The **Iráwadi** (*Iráwati*, the name of Indra's elephant) and the **Sittang** in Pegu, and the **Salwen**, the **Attaran**, and the **Tenasserim**, are the remaining rivers of British Burma (Barma). All are more or less navigable, and irrigate the country through which they pass; while innumerable tidal creeks give increased facilities of communication.

ISLANDS.—The islands are **Rámrí, Cheduba, Shapuri**, the **Mergui Archipelago**, and numberless others, to which must be added the **Andamáns** and **Nicobars**, now under the same Government. All the islands are very fertile.

146. *Climate.*—The climate is much the same in all parts at any given period of the year. The natives divide the year into the hot, the rainy, and the cold seasons, commencing in March, June, and October respectively.

[The maximum temperature is about 100° F., the minimum 60°, and the mean 80°. The annual rainfall is very considerable, varying from 60 inches, and even less in some places, to 190 inches in others. In March, April, and May the great heat partially or wholly dries up the rivers, and the surface of the country is everywhere parched; but in June, July, and August so much rain falls that not only are the channels of the rivers filled to overflowing, but the country around becomes so flooded that boats are the only vehicles which can be used, and ply everywhere over cultivated fields and open plains, and through forests, and streets of villages and towns.]

147. *Natural Productions.*—**Salt, coal, iron, tin, copper**, and **petroleum** are found in various parts. **Rice, tobacco, cotton, sugar**, and **pepper** are produced in great abundance in the low grounds; **timber**, chiefly

teak, is obtained in enormous quantities on the hills. The wild animals are of great variety and size; **game** is everywhere abundant; **tigers** frequent the forests, and **elephants**, which roam about in large herds, are constantly caught by the natives. **Sheep, oxen, buffaloes**, and **horses** are common.

148. *People.*—The people, who are called **Burmese**, slightly exceed **3,000,000** in number.

[This gives but a small average population to the square mile, but much of the country consists of mountainous tracts and impenetrable jungle, which has not even one inhabitant to the square mile. The inhabitants belong to various branches of the Indo-Chinese family, and have probably come down at a remote period from the plateau of Central Asia by the course of the Salwen and the eastern affluents of the Iráwadi. The most advanced race is the *Burmese*, which at the period of the British conquest possessed the ruling power over the whole country. The *Arakanese* are of the same race as the Burmese, have the same name, and speak the same tongue, but have been isolated for many generations by their geographical position. Hence they have local peculiarities of physiognomy and speech. The other races are the *Mon* or *Talain*, the ancient people of Pegu; the *Karen*, the *Khyeng*, and other hill tribes. All bear a general resemblance to each other.]

149. *Religion.*—The Burmese profess **Buddhism**, but the mountain races, except the Karens, who have been largely won over to Christianity, still retain the ancient worship of the deities of the woods, the hills, and the streams. They have no idols and no priesthood.

150. *Language.*—**Burmese** is the speech of three-fourths of the people.

[The hill tribes have dialects of their own, but they also speak

Burmese, and make use of it in their intercourse with strangers. Some of the hill dialects remain unwritten.]

151. *Education.*—The work of education is much more backward than in India, but is progressing. Village or monastic schools are very numerous, and there are a few Government schools.

152. *Industry.*—**Agriculture** is the chief pursuit, but it is carried on in a very rude manner.

[The hill tribes practise a method of tillage similar to that already described as in use amongst other barbarous tribes. When the soil becomes exhausted they remove their villages to other sites. In the large towns, which are the principal seats of industry, the making of common pottery, gold and silver work, the making of idols, and gilding and varnishing, are the chief industrial occupations.]

153. *Internal Communication.*—Water affords the chief means of communication in British Burma (Barma); there are no roads except those made by the British for military purposes.

[All the towns are situated on navigable rivers; and as the country is liable to periodical inundations (each over seven months in duration), generally speaking, ordinary roads would prove of little practical value. Steamers constantly ply between the chief ports and those of India. Telegraphic communication exists between the principal places and Calcutta. A railway, 163 miles in length, runs between Rangún and Prom.]

154. *Commerce.*—The commerce of British Burma (Barma) is carried on chiefly with Great Britain, Calcutta, Madras, Bombay, the Straits, and Ceylon. Its value exceeds ten crores of rupees. The *imports* are chiefly **cotton goods, hardware, machinery, woollen goods, silks, canvas,** and **tobacco.** The *exports* are principally **rice, timber, cutch, petroleum, hides, ivory,**

**cotton, gram,** and **ponies.** PORTS.—**Rangoon** (Rangún), **Moulmein** (Múlmaín), **Akyab, Tavoy, Mergui, Bassein** (Basín), and **Kyouk Hypoo** (Kyukfarú).

155. *Government.*—British Burma (Barma) is governed by a **Chief Commissioner** under the Government of India.

156. *Revenue.*—The revenue is derived from land, capitation-tax, fisheries, salt, forests, customs, excise, and stamps. It is increasing rapidly, and amounts to rather **more than sixteen lakhs of rupees annually.**

157. *Condition of the People.*—The social condition of the people is generally similar throughout the three divisions, and they have plenty of food and clothing. The houses of the peasantry, whether on the hills or on the plains, are built of bamboo, and have the floors raised on platforms so as to be above the reach of the annual floods; they are never built on the ground. The remote hill tribes still remain in a savage state of isolation and independence, but even the wildest grow cotton and weave cloth of strong texture and of various colours. All the tribes, as a general characteristic, are frank and truthful in the ordinary affairs of life, and they are also very hospitable.

158. *Towns in Arakan.*—**An** or **Aeng,** an important place on the route over the mountains to Ava, is reached from the sea by a navigable river. **Akyab,** on an island of the same name at the mouth of the Kuladan, is the most important town in the province, and well situated for commerce. **Kyouk Hypoo** (Kyukfarú), an important military station on the island of

Rámrí, is admirably situated for commerce, its harbour being one of the finest in the world.

159. *Towns in Pegu.*—**Bassein** (Basín), on the Bassein (Basín) river, a branch of the Iráwadi, has considerable trade, the river being here navigable for ships of the largest burthen. **Meaday**, on the Iráwadi, is a frontier town. **Pegu** stands on the Pegu river. **Prome** (Prom) is a place of importance on the Iráwadi. **Rangoon** (Rangún), the seat of government, and the most important town in Pegu, is situated on a branch of the Iráwadi called the Rangoon (Rangún) river, and has considerable trade; population, 100,000. Other important towns are **Henzada** and **Thyetmyo** (Tádítmoyo).

160. *Towns in Tenasserim.*—**Moulmein** (Múlmaín), the principal town in the province, situated on a small peninsula formed by the Salwen, the Gyne, and the Attaran rivers, is a fine port. It has some shipbuilding and great trade in timber. **Tavoy** and **Mergui** are other places of some importance.

---

## THE ANDAMA'N ISLANDS.

161. On the western side of the Bay of Bengal, and nearly south-west of Rangoon (Rangún) and Moulmein (Múlmaín), the volcanic group of the **Andamán Islands** extends nearly north and south along the 93d degree of east longitude, and between the 11th and 14th parallels of north latitude. The group consists of three large and several smaller islands, having a total area of about 1700 square miles. Dangerous coral reefs surround the whole, and the shores are fringed with belts

of mangrove. A dense tropical vegetation covers their whole surface.

[The inhabitants, in number probably not exceeding 2500, are at the very bottom of the scale of civilisation. They are about four feet eight inches high, and very black; have large teeth, and bodies tatooed all over. They have no government, and their religion consists in adoring the sun, the moon, the genii of the woods, and the spirit of the storms. They have no manufactures except bows, arrows, canoes scooped out of the trunks of trees, nets, paddles, twisted cord, wicker baskets, and thin knife-blades beaten out of old nails. They do not cultivate the soil, but wander from place to place, subsisting almost solely upon fish and the fruit of the mangrove. Their huts have no sides, and consist merely of roofs of palm-leaf thatch supported on four posts, from which are hung as ornaments bunches of fish and pigs' and tortoises' skulls. Owing to the difficulty of penetrating into the interior, the productions are imperfectly known, but many kinds of fruit and medicinal and timber trees grow in abundance. Hogs, rats, crows, and leeches are common, but these islands are otherwise deficient in animal life. The Andamáns afford many excellent harbours, the best of them being that of PORT BLAIR, on the east side of South Island. It is a magnificent landlocked bay, upwards of seven miles long, and varying in width from one to three miles. Port Blair is the penal colony for all India, the convicts being employed in making roads, felling timber, and otherwise bringing the island into a habitable state.]

---

162. **Barren Island** is little more than a volcano rising out of the sea to the eastward of the Andamáns. The island consists of a central cone 980 feet high, a circular valley, and an outside ridge, forming, as it were, walls to keep out the sea. The cone occasionally sends out showers of red-hot stones and enormous volumes of sulphurous vapour.

---

163. **The Nicobar Islands.**—These islands lie to the south of the Andamáns. Traffic is carried on with

some of them in cocoa-nuts, betel-nut, pigs, poultry, and yams.

[The natives, like the Malays, to which race they belong, are piratical, but are, compared with the Andamánese, perfectly civilised. They wear clothing, speak a little English, and construct their boats and huts with great ingenuity. The Danes formed a settlement on this group in 1756, but abandoned all right to the islands in 1846 in consequence of their unhealthiness.]

---

## BHUTÁN.

164. *Boundaries.*—Bhután (the ancient *Madra*) is bounded on the north by the main chain of the Himálayas, on the east by tracts of country occupied by rude mountain tribes, on the south by Assam (Asam) and Kuch Bihár, and on the west by Sikkim and Dárjiling.

165. *Physical Features.*—Towards the south another range, about 8000 feet in height, with occasional peaks rising to an altitude of 16,000 feet, runs parallel to the stupendous snow-clad Himálayas, and farther south again is a third and smaller parallel range. Between the Himálayas and the second range is a high tableland, much too bleak and barren to be habitable, except at the foot of the latter range, where most of the principal places in the country are situated. Beyond the second range, in an eastern direction, the land is level, and southward of the third range, which is of very considerable height, are the **Duárs**, tracts of country of surpassing fertility, the produce of which forms the principal means of subsistence of the people.

[Bhután presents a variety of climate and scenery scarcely to be equalled by any other country in the world. The cold of Siberia, the heat of Africa, and the pleasant warmth of Italy may all be experienced in a single day's journey; and at one view may be seen rugged barren hills and valleys covered with luxuriant vegetation, rushing mountain torrents and gentle streams, dense forests and sunny slopes, placid lakes, steep precipices, and vast masses of eternal snow stretching upwards till they are lost to sight.]

166. *Productions.* — Bhután produces **rice** and **millet**; **sheep, ponies**, and a small and hardy breed of **horned cattle** are reared; and **game** of all kinds abounds in the forests.

167. *Commerce.*—There is almost no trade for want of roads. The ordinary routes are mere tracks, and when in the rainy season the ravines through which they often pass become the beds of considerable streams, the natives travel along small and irregular pathways traced out on the hill-sides.

168. *People, &c.*—The Bhutánese are divided into three classes—priests, Government servants, and cultivators. The priests, who are very numerous, live in castles and the best houses, and occupy the best villages. Their chief business is to be idle, to feast at the expense of the people, to count their beads and mutter prayers. They alone possess any education; the other classes are totally ignorant. The cultivators are very poor, and live in miserable huts, and the Penlows or chiefs and the governing classes are but little more luxurious than the ordinary people. Women are treated as little better than beasts of burden. Nominally the country is governed by a temporal and spiritual head called the Dharm Rája, supposed by the Bhutánese to be a divinity in human shape; but the affairs of state are really conducted by a functionary styled the Deb Rája, who is elected by the Penlows every

three years from among themselves. The whole system of government, however, is so bad that there is no security for property, and not much for life. Owing to this, and to the filthy and immoral habits of the people, the population is scanty, and the villages are few and small. The Bhutánese profess Buddhism, and their language, which consists of many dialects, is called Bhutía.

169. *Towns.*—**Panakha** is the capital and seat of Government, and generally the place of most importance in Bhután. **Tasi-chozong** and **Toungsu** are other places of local importance.

---

## NEPÁL.

170. *Physical Features.* — **Nepál** (the Sanskrit *Népála Désa*) stretches from the valley of the Ganges on the south, to the crest of the main chain of the Himálayas on the north, and from Sikkim on the east to Kumaun on the west. It is about 55,000 square miles in area. The country is traversed by the Gandak (Gandakí), the Kusí, and several other considerable streams, and is naturally divided into five parallel zones. The Tarái extends all along its southern border and for about ten miles northward; a true forest region, producing a great variety of valuable timber, then begins, and stretches northwards for another ten miles. Beyond this the country becomes hilly, and rises in terraces gradually towards the north, till the surface breaks into mountains, forming a belt thirty miles wide, beyond and above which towers the

great snowy range with **Mount Everest,*** **Daulagiri, Kanchinjinga,** and others—**the highest peaks in the world.** In the mountainous region several thickly inhabited valleys occur, running in a direction north and south, and varying in height from 3000 to 6000 feet above the plains of Bengal. That of Nepál proper, the largest, is of oval shape, twelve miles long and nine broad. It is bounded on all sides by lofty mountains, and its wave-like surface is covered with a rich expanse of cultivation, watered by numerous winding streams and studded with villages and towns. From its appearance it is supposed that at one time the valley of Nepál was a lake.

171. *Climate.*—The climate of Nepál is characterised by great extremes in different parts. All degrees of temperature, from the heat of the Bengal plains to the cold of Siberia, may be experienced in a day or two's journey. As in Upper India generally, the rains commence rather earlier than in Bengal.

172. *Products.*—**Copper** and **iron** are wrought in the hills, **building stone** is plentiful, and **arsenic** and **lead** occur. Some of the valleys produce **rattans** and **bamboos** of enormous size; others nothing but **oaks** and **pines**; others **sugar-cane** and **fruits**; and others **rice, wheat, millet**, and similar grains. **Ginger** and **cardamoms** abound, and **oranges** grow to perfection. Everywhere and at all seasons the ground teems with the most beautiful flowers in great variety. The domestic animals of Nepál are as a whole very inferior, but

---

* *Mount Everest* is called after Sir George Everest, a British officer, at one time head of the great Trigonometrical Survey of India.

large flocks of **sheep** of a valuable **kind** are reared on the mountains. Wild animals and **game** of all sorts abound; herds of **elephants** range in the forests.

173. *People, Religion.*—The population, which amounts to two millions, consists of **Ghurkas, Newars, Bhutías** (Tibetans), and **aboriginal mountain tribes.** The Ghurkas, the ruling race, are Hindus by descent and religion: the Newars are agriculturists, traders, and artisans; they have Chinese features and are of Tartar origin. The Bhutías live far up the mountains adjoining Bhot or Tibet. They and the Newars are Buddhists in religion.

[The food of the poorer classes consists of rice, garlic, and lentils. All ranks drink intoxicating liquors to excess, and the people in the eastern part of the country are described as spending their nights in debauchery, their mornings in sleep, and their days in the performance of religious ceremonies. Throughout Nepál most of the domestic servants are slaves.]

174. *Language.*—The language spoken by the Hindus of Nepál is called **Párvatíya** (Sans. *párvatíya* = belonging to a mountain); that of the Newars is peculiar to themselves.

175. *Industry.*—The manufactures include **utensils of copper, brass,** and **iron**; **casting of bells, weaving of coarse cotton cloths,** and **making of paper. Cotton goods, raw cotton, raw silk, gold** and **silver lace, carpets, cutlery,** and **tobacco,** are imported from Bengal and Oudh; and **sheep, musk, quicksilver, borax, drugs,** and **Chinese silks** from Tibet. The exports are **elephants, rice, timber, hides, iron, copper, ginger, honey,** and **fruits.**

176. *Towns.*—**Katmandu,** the capital, is situated in

the Nepál valley. The rája resides here, as also does the British Resident. **Lalitapatan** (= Town of Delights), **Ghurka**, and **Makwanpur** are other towns of some importance.

[Nepál is under the government of an independent rája. Formerly it was possessed by numerous rulers, and extended westward as far as the Satlej. About the middle of the last century, the Rája of Ghurka conquered the territories of the other rájas, one by one, until he acquired the whole of the country. Depredations were then committed in British territory; war ensued, and as its result all the country west of the river Kálí was added to the British dominions.]

---

## THE NORTH-WEST PROVINCES OF BENGAL AND OUDH.

177. These provinces stretch from Bengal proper on the east to the Panjáb on the west, and from Bundelkhand, Central India, and Rájputána on the south, to Nepál and Tibet on the north. Their estimated area, exclusive of Oudh, is about 80,000 square miles.

178. *Physical Features.*—The whole country is physically divisible into the Tarái on the north, the great valley of the Ganges in the centre, and the northern portion of Bundelkhand on the south. The Tarái is a narrow strip of country lying at the foot of the hills. It is a pestilential tract, abounding in swamps and impenetrable jungle, and infested with fierce wild beasts. Large portions have, however, of late years been brought under cultivation. The valley of the Ganges presents one vast plain rising imperceptibly from east to west; at Allahábád the general surface of the country is about 400 feet above the sea, and at Sahárampur it is about 1000 feet above the same

level. In travelling towards the west the soil becomes more and more sandy and arid; in the east wells are merely aids to cultivation, in the west they are absolutely necessary: at one time this great tract teems with fertility, and at another suffers from drought; hence the construction of costly canals in the Duáb, which wants water only to make it the garden of the world. The Bundelkhand districts rise southward from the plains, and the surface of the country gradually grows more and more uneven till it breaks into the rocky hills of Bundelkhand proper.

[West of the Ganges the soil contains great quantities of *kankar*, an inferior kind of lime, which is generally found in irregular masses, resembling the roots of ginger. The lumps vary in size from pieces weighing five or six pounds to others scarcely an ounce. Sometimes kankar appears in enormous beds. All the streams which join the Ganges on the right bank are more or less impregnated with this kind of lime; while on the opposite bank the waters contain in solution great quantities of nitre, a mineral with which the plains of that part of India abound. From these causes many streams on both sides of the Ganges not only afford no aid to the cultivator, but greatly injure vegetation, and serious sicknesses attack the people, who, either through ignorance or indifference, drink much of the waters.]

179. *Climate.*—The climate is subject to great extremes. In the cold season the temperature is often such as will freeze water, but in the hot season dreadfully oppressive.

[About the middle of April, hot westerly winds, sweeping over the deserts of Rájputána and Sindh, begin to blow with fury. The districts nearest the west are of course most affected. The wind rises regularly about eight in the morning, and, continuing to blow till sunset, lasts thus all through May and into June, increasing in intensity. The country loses its verdant aspect, vegetation withers, and clouds of penetrating dust fill the air. Every house is shut closely up, and is little better than an oven. About the middle of June the dust grows denser, the thunder roars, the lightning flashes through the gloom, and the rain pours down in torrents, flooding the country, and giving relief to all animal life. With the first fall of rain the country is overspread as with a mantle of verdure, and from October to March the weather is dry, cool, and delightful.]

180. *Natural Productions.*—These provinces, like Bengal generally, are almost destitute of mineral wealth. **Iron** in considerable quantity is found in Kumaun, as also is **copper**; **nitre** is obtained from the soil of the plains; and **sandstone** for building from Chunár. **Cotton** is the chief commercial product, and is grown extensively in every district. **Rice, maize, millet, indigo, wheat, barley, oil-seeds,** and **tobacco** are largely cultivated, and fruits and vegetables of all kinds are abundant. The hilly districts afford large supplies of **timber. Tea** of excellent quality is grown in Dehra Dún, Kumaun, and Garhwal. The Tarái abounds with all kinds of wild animals. The domestic animals are inferior.

181. *People, Religion.*—The population is estimated at 30,000,000. The prevalent religions are **Hinduism** and **Muhammadanism.** The temples and holy places of the former are more numerous in these provinces than in any other part of India. It is estimated that more than two million pilgrims visit Hardwár alone in the course of the year.

182. *Language.*—The common vernacular is **Hindí.** *Hindí* and *Urdú* are taught in the schools.

183. *Education.*—The annual sum expended by the state on education is about fourteen lakhs of rupees.

184. *Industry.*—The people are almost wholly engaged in **agriculture.** **Indigo** is manufactured in many districts; **weaving, pottery,** and some kinds of ornamental work are carried on in towns.

185. *Internal Communication.*—The roads are many and excellent, and water-carriage is abundant in almost every district. In the canals of the Ganges and Duáb and their offsets, these provinces possess some of the finest works of the kind. A railway runs through the heart of the country from Benares (Banáras) to Dehli.

186. *Commerce.*—**Shawls** and **other woollen fabrics** are brought from Cashmere; **camels, horses, mules, salt, antimony, fine woollen** and **cotton goods, asafœtida,** and **dried fruits** from countries west of the Indus; **salt** and **fancy goods** from Rájputána; and **calicoes** from Calcutta. The principal exports are **cotton** and **indigo,** which are sent by water to Calcutta.

187. *Revenue.*—The revenue amounts to about **five crores of rupees annually,** and is chiefly derived from land, abkaree, and stamps.

188. *Government.*—The Government of these provinces is administered by a **Lieutenant-Governor.**

189. *Condition of the People.*—The common people are very poor and ignorant. Their food consists chiefly of chupatties or unleavened wheaten cakes; their dress is somewhat fuller than that of the people of the Lower Provinces; their dwellings are mud cottages

like those of Bengal, but covered with tiles instead of thatch.

190. *Divisions.*—The divisions are seven in number: 1. **Meerut** (Mírath), comprising the districts of Dehra Dún, Sahāranpur, Muzaffarnagar, Meerut (Mírath), Bulandshahr, and Aligarh; 2. **Kumaun**, including Kumaun and Garhwal; 3. **Rohilkhand**, comprising Bijnor (Bijnaur), Muradábád, Budaun, Bareilly (Barelí), Sháhjahanpur, and Tarái; 4. **Ágra**, comprising Muttra (Mathurá), Ágra, Farrukhábád, Mainpurí, Etáwah, and Etah; 5. **Jhánsi**, comprising Jalaun, Jhánsi, and Lalitpur; 6. **Allahábád**, comprising Cawnpore (Kánhpur), Fathipur, Allahábád, Bánda, Hamirpur, and Jaunpur; and 7. **Benares** (Banáras), comprising Gorakhpur, Bastí, Azamgarh, Mírzápur, Benares (Banáras), and Gh).

191. **Meerut** (Mírath) **Division.**—The surface of this division slopes gently towards the south, and is remarkably uniform throughout. It is watered by the Ganges and the Jamna, and numerous torrents which rush down from the Sewálik Hills into those rivers; and is artificially irrigated by the Duáb and Ganges canals. The soil is mostly clay mixed with sand; it produces wheat, barley, oats, pulses, tobacco, and European vegetables in the cold season, and rice, cotton, indigo, maize, and millet in the hot. Hemp grows wild in great luxuriance.

**[Along the banks of the Jamna and Ganges there are strips of marshy land called Khádar, very fertile, and well adapted for growing rice, but unwholesome, and consequently thinly inhabited and poorly cultivated.]**

*Towns.*—**Sahāranpur** is the principal town. Being situated nearly on the northern limit of the flora of India, the Government have here established botanical gardens, where the plants and trees of all countries may be seen growing in the open air. **Rúrkí** stands on one of the highest sites in the Duáb, and on this account the Ganges canal is made to pass here, whence its waters can be directed to all the lower lands. Rúrkí has an important civil engineering college, and an immense aqueduct, two and a half miles long, across the Soláni valley. **Haridvár**, called also Gangá-dvára (= the Door of the Ganges), is a famous place of pilgrimage. At a little distance is the head of the Ganges canal. **Meerut** (Mírath) is the chief town in its collectorate. The revolt of the Bengal sepoys commenced here on the 10th May 1857.

[The districts of Saháranpur, Muzaffarnagar, Meerut (Mírath), Bulandshahr, and Aligarh were ceded to the British by Sindia, under the treaty of Serji Anjengaon, December 1803.]

192. **Dehra Dún.**—Dehra Dún (= Valley of Dehra), forming a part of the Meerut (Mírath) division, is a valley nearly 700 square miles in extent, almost surrounded by hills, varying from 3000 to 8000 feet in height, and separated from the district of Saháranpur by the Sewálik mountains.

[A slightly elevated ridge crosses the valley from south-east to north-west, and divides it into two basins, the streams on one side flowing into the Ganges, and those on the other into the Jamna. The temperature of the valley varies from 37° to 101° F., the mean being 70°. During the rains the climate is very unhealthy, and fevers are common. The soil is a deep, rich mould, producing rice, maize, gram, sugar-cane, cotton, hemp, opium, and indigo; and apples, pears, plums, apricots, and other European and Indian fruits come to perfection. Tea of excellent quality is also grown in considerable quantity; and the uncultivated portions are covered with dense jungles, abounding in wild animals and feathered game.]

*Towns.*—**Dehra**, the chief town, stands on the elevated ridge which crosses the valley from south-east to north-west. **Mussoorie** (Masúrí), 7000, and **Landaur**, 8000 feet above the sea, are situated almost close together on the northern side of the Dún. They are sanitaria for Europeans, and the latter is a depôt for sick soldiers.

[There is no level ground at either place, and most of the houses are built upon terraces cut out of the solid rock. The views from these stations are very beautiful. On the north, the eye rests upon successive tiers of mountain ranges, terminating in the snowy peaks of the Himálaya; on the south, the Dehra Dún, more than 4000 feet below, appears with its fields, forests, and rivulets; and beyond the Sewálik range, as far as the vision can reach, are seen the fertile plains of Upper India.

The Dehra Dún was wrested from the Ghurkas by the British in 1815.]

193. **Kumaun Division.**—The division of Kumaun includes the districts of Kumaun and Garhwal.

[The southern portion is a dense woody region called the Báhbar: it is destitute of streams, but infested with wild beasts. Northwards the country breaks into rugged mountain-masses rising into peaks, some of which are perpetually white with snow. Vast tracts of this province are composed of rocks or covered with forests; in some parts such features are confined to the upper parts of the mountains, whose sides and base are adorned with the greatest fertility; while, in other parts, the finest slopes or the fairest valleys are succeeded by miles of narrow ravines, where precipices and woods extend to their lowest depths. Sometimes the jungle is below the cultivated tracts, sometimes above. The rivers are very numerous, and are all feeders of the Ganges. Many parts of Kumaun are very beautiful; the mixture of the natural scenery of wood and water, the care-displaying fertility of the fields, and the hills dotted with the habitations of the people, make up a picture which it would be difficult to equal in any other part of India. Gold-dust is occasionally found in the streams; copper is obtained in considerable quantity; and iron of fine quality abounds. The vegetation varies much according to elevation, climate, and the nature of the soil. In the upper regions no more than one harvest of wheat, barley, buck-wheat, and turnips is possible; and in some years even this is lost by early falls of snow. Lower down, two harvests a year are reaped; rice, cotton, indigo, maize, ginger, and turmeric form the chief crops of summer; and wheat, barley, tobacco, flax, and European vegetables those of winter. Hemp is grown too, and much tea of excellent quality. Besides the forests of pine, cedar, sâl, and oak, there are many varieties of fruit trees, such as the walnut, the orange, and the pomegranate. From the fruit of the latter a fine acid is made, and its rind is used in tanning. Elephants are numerous in the Báhbar; bears and leopards are also common, and tigers, found far up the mountains, prowl about the open country, and very frequently carry off men as well as cattle. On the higher mountains, near the snow, large flocks of sheep and goats are bred and pastured for the sake of their wool, and as beasts of burden. Where these animals are not used, men laboriously carry the produce of their fields on their backs to market. All the routes are exceedingly rugged, and those over the mountains into Tibet are rendered so difficult and dangerous by snow-beds, precipices, and waterfalls, as to be seemingly impracticable. The carrying trade is in the hands of Tibetans who occupy the villages in the dreary elevated regions. They are most active and enterprising, and exceedingly jealous of their right to monopolise their dangerous business. Grain, sugar, spices, broad-cloths, calicoes, and tobacco are taken from the south into Tibet, in exchange for shawls, goats' and sheep's wool, borax, leather, dried fruits, yaks' tails, and drugs. The commerce of the lower country consists in sending the products of Kumaun to Oudh chiefly, in exchange for the manufactures of Britain and India.]

194. *People.*—The people are mostly **Hindus**.

[In the more elevated parts they wear woollen garments, and in the warmer valleys to the south clothes of cotton or hemp. Some people do not change their clothes all the year round; they are consequently subject to severe sicknesses; and in some parts, where the habits of the population are equally filthy, especially in the neighbourhood of copper mines, many are cretins or mountain idiots; while others, and frequently the cretins also, are afflicted with goitre, an enormous tumour which forms on the fore-part of the neck.]

195. *Religion.*—The religion is **Bráhmanism.**

[Almost every peak, forest, rock, and spring in the country has its presiding deity, to whom regular offerings are made; while the numerous prayágs, or forks of rivers, are considered more than ordinarily sacred, and Kedárnáth, Badrináth, and other most holy shrines of the Bráhmans, are visited from May to October by large bodies of pilgrims.]

196. *Towns.* — **Almora,** the chief town, stands almost in the centre of Kumaun, on a ridge of hills in the midst of a beautiful and highly cultivated country. There are inexhaustible quantities of stone and slate in the neighbourhood. **Naini Tal** is the sanitarium of the province. It is situated 7000 feet above the sea, on the border of a beautiful lake from which it takes its name. **Srínagar,** the former capital of the country, is now a decayed place.

197. **Rohilkhand Division.** — Rohilkhand (= the Country of the Rohillas) extends from where the Ganges leaves the hills to the frontier of Oudh, and from that great river on the south northward to the hill districts of Kumaun. It thus lies between the Duáb and the mountains. In all its principal physical characteristics it closely resembles the Duáb. Its surface is flat, sloping slightly and gradually to the south-east; and its southern portion is subject to annual inundation. The soil is very fertile, and produces **rice, maize, wheat, cotton, sugar, tobacco,** European vegetables, **plantains, dates, walnuts,**

**grapes, strawberries,** and many other kinds of fruit in profusion.

[As a consequence of the slight inclination of the country, the numerous small streams which trickle from the hills stagnate at their foot, and, saturating the fertile soil, give growth to gigantic trees, creepers, and underwood, forming an impenetrable jungle.]

*Towns.*—**Bijnor** (Bijnaur), **Moradábád** (Múradábád), **Budáun, Bareilly** (Barelí), **Sháhjahanpur** are the chief towns in their respective districts. Bareilly (Barelí) is the principal town in Rohilkhand.

198. **Ágra Division.**—The districts of this division lie for the most part in the Duáb, but partly also south of the Jamna. The whole of the tract is flat, sloping gently to the south-east. Where irrigated, the soil produces luxuriant crops of **rice, wheat,** and other grains grown in the Duáb; but in other parts the land presents a miserably arid appearance.

*Towns.* — **Muttra** (Mathurá), **Ágra, Mainpurí, Etáwah, Kásganj,** and **Farrukhábád,** are the principal towns in their respective districts. **Muttra** (Mathurá), situated on the Jamna, is regarded as sacred in Hindu mythology from being the birthplace of Krishna, the incarnation of Vishnu. The town was plundered in A.D. 1017 by Mahmud of Ghazni. **Ágra,** on the south bank of the Jamna, was formerly the favourite residence of the Mughal emperors. In the neighbourhood is the world-renowned Táj Máhal. Near it also is one of the finest buildings in India—the tomb of the Emperor Akbar.

[Muttra (Mathurá), Ágra, and Mainpurí were ceded to the British by the treaty of Serji Anjengaon in December 1803;

Rohilkhand, Etáwah, and Farrukhábád became British territory in 1801 by an arrangement with the Nawáb of Oudh.]

199. **Jhansí Division.**—The small tract comprehended by this division lies south-east of the last. Its chief towns are **Kalpí, Mau,** and **Lalitpur.**

200. **Allahábád Division.**—This division lies partly in the Duáb and partly south of the Jamna. The country included in the former bears all the characteristic features of the Duáb—a plain sloping gradually south-eastward, arid where destitute of water, but where irrigated wonderfully fertile. The tract to the south of the Jamna is flat for some distance from the river, and then breaks into undulating ground, forming a transition between the plains and the hilly region of Bundelkhand. Cotton is very largely cultivated, as also are maize, millet, and indigo. This province possesses great facilities of communication in its fine navigable rivers, good roads, and the railway.

*Towns.* — **Cawnpore** (Kánhpur = Krishna's City), **Fathipur, Allahábád, Bánda, Ráth,** and **Jaunpur** are the principal places in their respective districts. **Cawnpore** (Kánhpur) stands on the right bank of the Ganges, here five hundred yards wide. It has a large cantonment, and a very flourishing trade. In the mutiny of the Bengal sepoys in 1857 all the principal buildings were destroyod, and a number of English women and children barbarously massacred. Population, 114,000. **Allahábád,*** the *Váranávata* of the

* "The abode of God." Hindus regard it as holy, while the superstitious believe that a third river, the Saraswati, runs underground, here meeting the Ganges and Jamna, and that bathing in the united stream with the head newly shaved, so that each

"Mahábhárata," the seat of the Government, is a large town at the confluence of the Ganges and Jamna, each here nearly a mile in width. Allahábád is not noted for art or manufactures of any kind, but it is a great place of commerce. Population, 105,000. **Bánda** is a thriving place, and a great mart for cotton.

[Cawnpore (Kánhpur), Fathipur, and Allahábád were ceded to the British in 1801 by the Nawáb of Oudh. Bánda and Hamirpur came into British possession in 1802 by treaty with the Péshwa.]

201. **Benares** (Banáras) **Division.**—The tract of country which constitutes the Benares (Banáras) Division, stretches from the Nepál Tarái on the north to the Bundelkhand plateau, including a part of it on the south. With the exception of the southern portion of Mírzápur, which is elevated and jungly, the whole province is flat. The Gogra, the Gumtí (Gomatí), the Ganges, the Són, and many smaller streams, afford it great facilities for transport. Great numbers of small lakes are scattered over the surface, and the soil is very productive.

*Towns.*—**Mírzápur, Benares** (Banáras), **Gházípur, Azamgarh,** and **Gorakpur,** are the principal towns in their respective districts. **Mírzápur** is famed for its carpets. **Chunar** (Chanár), on the right bank of the Ganges, has a fort in which Trimbakji Dainglia was kept captive. **Benares** (Banáras)—the ancient *Váránasí** or

---

hair may fall into the water, gains for the devotee a million years' residence in heaven for every hair so falling. At certain times of the year, therefore, many thousands of people flock to this city to go through the ceremony.

* From *Vara* and *Nasi*—the names of two rivers.

"the Lotus of the World"—is situated on the left bank of the Ganges. It extends for about three miles

River-side scene, Benares.

along the river, and fine gháts or flights of stone steps lead up from the river to numberless temples, mosques, and palaces, which rise in rows one above another. The ordinary houses are two storeys high, but many of them are six storeys in height. All the houses have exceedingly small windows, and many of them have their upper stories connected by balconies with those on the opposite side of the street. The streets themselves are so narrow, crooked, and crowded, that it is difficult to pass through them on horseback; and most of the business of the place is transacted on the gháts. There, at all times of the day, may be seen thousands of people bathing, praying, preaching, gossiping, bargaining, lounging, and sleeping. Pilgrims from all parts of India visit Benares (Banáras), and it is said that there are 8000 houses occupied by Bráhmans, who subsist on alms and offerings; while in every part of the city crowds of fakirs may be met with in all the hideousness of filth, cow-dung, disease, and deformity. Benares (Banáras) is a place of much trade, and many of its merchants are exceedingly rich. Pop. 175,000. **Gházípur**, on the Ganges, is a large town, chiefly noted for its roses, large quantities of which are grown in the neighbourhood for making rose-water and attar. Lord Cornwallis died here in 1805.

---

## OUDH.

202. *Physical Features.*—Oudh is situated in the centre of the great sub-Himálayan valley drained by the Ganges and its tributaries. An irregular line running from the Ganges near Fathigarh to the Pilibhit

Tarái, above Kháiragarh, separates the province from Rohilkhand on the west; a similar line, extending from the Ganges across the Gogra to the Nepál hills, forms its eastern limits; on the north it is bounded by the lower ranges of the Himálayas, and on the south by the Ganges. On the map Oudh lies obliquely in the form of a parallelogram, with a superficial area of 24,000 square miles. Its general surface may be regarded as a plain sloping, as the direction of its rivers indicates, from north-west to south-east. The **Kauriáli** or **Gogra** (the ancient *Sarayú*), the principal river, is navigable for steamers to Faizábád; the **Gumti** (Gomatí = winding), though navigable, is of little commercial importance on account of its winding course. Most of the other streams are mere rivulets, except the **Raptí**, which is a mountain torrent.

203. *Climate.*—Like that of other parts of India in the same latitude, the climate of Oudh is dry during the greater part of the year, and subject to wide extremes of temperature, the thermometer sometimes indicating 112° F., and at other times sinking below freezing-point.

[The seasons are divided into the cold, the hot, and the wet. The cold weather lasts from the end of October to the beginning of March; the hot from that time to the middle of June, and the wet from thence to the end of October. During the hot weather disagreeable sultry winds laden with dust blow from the west, or occasionally shift round to the east, bringing with them oppressive vapours from the swamps of Bengal. Sometimes hurricanes occur and do extensive damage. The annual rainfall varies from eighty to thirty inches.]

204. *Products.*—Oudh is so fertile that few portions of India surpass it in productiveness. The staple

products are **wheat, barley, maize, rice, oil-seeds, sugar-cane, indigo, opium,** and **cotton, salt, saltpetre,** and **soda.** Valuable **timber** is plentiful in the forests. **Elephants, bears, wild hogs,** and **deer** swarm in the Tarái; and **tigers, wolves, hyenas, jackals,** and hosts of smaller animals abound in the open country.

205. *People, Religion.*—The population, estimated at eleven millions, consists of **Hindus** and **Muhammadans.** The Hindus greatly predominate.

206. *Commerce.*—The trade of Oudh is very considerable; **European stores** of all sorts, **piece-goods, yarn,** and **hardware** are imported from Calcutta; **salt** and **cotton** from the adjacent districts; **cattle, copper, iron,** and **spices** from Nepál. The *exports* are **grain** of all kinds, **linseed** and other oil-seeds, **opium, ghee, hides, saltpetre, lac, catechu, timber,** and **cloth** of Tánda manufacture, chiefly to Nepál. The inland trade is assisted by annual fairs, the largest of which is held at Khairábád.

207. *Towns.*—**Lucknow** (Lakhnau), the chief town, is advantageously situated on the right bank of the Gumti (Gomatí). A branch line of railway connects it with Cawnpore (Kánhpur). **Oudh** (the ancient *Ayodhyá*), the former capital, stands on both banks of the Gogra; it is much venerated by the Hindus, and contains many ruins of great antiquity. Adjoining it is **Faizábád,** a commercial town with manufactures of cloth, metal vessels, and arms. **Rái Barelí** is the chief place in the division of the same name. **Bah-**

raich, Partábgarh, Gonda, and Sítápur are other towns of some importance.

[Oudh is the ancient *Kosala.* Its capital, Ayodhyá, was the birthplace and seat of a race of kings who boasted their descent from the sun and moon, and became the progenitors of the princes of all other countries in India. The far-famed Ráma was its king two or three centuries before Christ. After him sixty princes ruled, and the seat of government was transferred to Kanuj. The boundaries of the new kingdom extended as far as the Chambal and to Ajmír, and were maintained till A.D. 1193, when the Hindu dynasty was subverted by the Muhammadans. It then became part of the Dehli empire, and under Akbar was constituted a Subáh governed by a viceroy. Thus it remained until the dissolution of the Mughal empire, when its rulers rendered themselves independent, and transmitted their authority to their heirs. The founder of the dynasty, Sa'adat Khán Buran-ul-Mulk, originally a merchant of Khorassan, rose to high military command in the reign of Muhammad Sháh, and attained to such power in his province, that he not only repelled the attack of a powerful enemy, but marched with a large force to the aid of Muhammad when India was invaded by Nádir Sháh. He died in 1739, and Oudh remained in the possession of his descendants down to 1856. For many years previous to the latter date, however, the country had been allowed to fall into the most frightful disorder. Its princes one after the other wholly neglected affairs of state; the law was openly defied; rich men made fortresses and lived in them for protection, while poor men went about armed to the teeth for security. The army subsisted by plunder; no property was safe; and the millions of Oudh were for years strangers to everything but bloodshed, oppression, and misery. The British Government had long and repeatedly remonstrated with the Court of Oudh, but to no purpose; the treaties which had been made were of no avail, and nothing remained but to place the country under British rule. This was done in February 1856. The king and his relatives were pensioned; the fortresses which covered the country were demolished, the people disarmed, and law and order restored. Roads have since been made, bridges built, and police, jails, hospitals, and schools established; and Oudh is now becoming in every respect one of the foremost parts of India.]

## MÁLWA, OR CENTRAL INDIA.

208. **Central India** is an irregular tableland, varying in elevation from 1500 to 2500 feet above the sea, of vast extent, well watered, and exceedingly fertile, and possessing one of the mildest climates in India. It is divided into a number of principalities held by native chiefs, the principal being **Indor** (Indaur), or Holkar's dominions, **Gwáliár**, or Sindia's dominions, **Bhopál**, **Dhár**, and **Ratlam**. The states of Bundelkhand are also included in Central India. All the territories are under the political superintendence of the Viceroy and Governor-General's Agent, who resides at Indor (Indaur).

209. **The Indor** (Indaur) **Territories.**—These territories consist of several widely scattered districts, covering altogether more than 8000 square miles. The northern parts are watered by the Chambal and its tributaries; the southern portions form part of the valley of the Narbada (Narmadá), and are traversed by the Vindhya and the Sátpura mountains. All the districts are very fertile, and produce wheat and other grains, opium, sugar-cane, tobacco, and cotton in abundance. Opium is so largely grown, that when the poppy is in bloom the country looks like a flower-garden. The inhabitants are chiefly Maráthas and other Hindus, Muhammadans, and a considerable number of Bhíls, estimated altogether at about a million.

*Towns.*—**Indor** (Indaur), the capital, is the chief town of Málwa, and the station of the British Resident. Fourteen miles south-west of Indor (Indaur) is **Mhau**, a large military station. It is upwards of 2000 feet above the sea, very healthy, and situated in the midst of beautiful scenery.

[The family of Holkar was founded by Malhar Ráo Holkar, born in 1693, the son of a cultivator in an obscure village of the Dakhan. He was at first a shepherd, but entering the service of the Péshwa as a soldier, he rose rapidly to distinction, and was appointed to the command of five hundred horse. Soon afterwards he received a jághir of twelve districts north of the Narbada (Narmadá), next seventy more, and two years subsequently Indaur and its possessions were added. Holkar then became one of the chief Márátha leaders, and fought as such at the disastrous battle of Panipat in 1761. He died, and his successors came into repeated collision with the British, until hostilities were terminated by the decisive victory gained by the latter in 1817 at Máhidpur. Holkar then agreed to the usual conditions of renouncing all diplomatic intercourse with other states, and to entertain no foreigners without the consent of the British; to discharge his superfluous troops, and to receive a British Resident. On the other hand, the British engaged to maintain a force for the preservation of internal tranquillity, and for defence against foreign aggression. The Indaur territory is governed by a Mahárája.]

210. The **Gwáliár Territories.**—These territories are of the most irregular form, consisting of several detached districts, covering a total area of more than 33,000 square miles. They are remarkably well watered by the Chambal, the Taptí, and the Narbada (Narmadá), and their numerous tributary streams. Except in the northern parts, where the rocky sandy soil and torrid climate resemble those of Rájputána, the climate is mild and equable, the soil is fertile, and the productions are similar to those of Indor (Indaur). The population exceeds 3,000,000, and consists of Máráthas, Játs, and Rájputs. Bráhmans are numerous.

*Towns.*—**Gwáliár**, the capital, is a town of 50,000 inhabitants. The celebrated fortress of Gwáliár is built on a long flat isolated hill, half a mile in length, three or four hundred yards broad, and rising in the highest part 450 feet above the plain. The works are so extensive that 15,000 men would be required to adequately defend them. It is stated to have been built before the commencement of the Christian era, and amongst its captors may be named Mahmud of Ghazni, Bábar, Akbar, and the British. Akbar made it a state prison, and in it Aurangzeb first confined and afterwards put to death his brother Murad and his son, and the two sons of Dára. **Ujjain** (Ujjaiyiní) contains a great number of Hindu temples. The ancient city of Ujjain (Ujjaiyiní), now in ruins, was the capital of Málwa, and one of the seven sacred Hindu cities. It was also the first meridian of Hindu geographers. Several of its kings were very famous; and B.C. 57, the year in which one of them (Vikramáditya) began his reign, forms an era in Hindu chronology. **Bhílsa** is a considerable town. In its neighbourhood are many curious Buddhist remains, and the tobacco raised in the vicinity is considered the finest in India. **Mandisur** is on a tributary of the Chambal. Here in 1818 the final treaty between Holkar and the British was concluded.

[Sindia, the founder of the ruling family of Gwáliár, like Holkar, rose from a very humble station to be one of the chief of Márátha leaders, and Mádhaji Sindia, his son and successor, and the principal opponent of the English in the first Márátha war, attained to still greater power. At his death in 1794 his territories included Khándesh, the greatest part of Málwa, part of the Dakhan, Ágra, Dehli, and the finest parts of the Duáb.

Daulat Ráo Sindia, his successor, having joined in the Márátha invasions of the Nizám's dominions, was defeated at Assaye in September 1803, and compelled to relinquish all his possessions north of the Chambal. Daulat Ráo died childless in 1827, and was succeeded by Ali Jáh Jankoji Sindia, who also died childless in 1843, and was succeeded by Ali Jáh Jaiaji Sindia, a boy of eight years of age. His regent was overthrown by military violence, and the government fell into the hands of the widow of the late Mahárája, a capricious girl, twelve years old. Bloodshed and anarchy ensued, and towards the end of December 1843 a British force marched into Gwáliár to restore order. Two obstinate engagements followed, in both of which the Márāthas were defeated, and a treaty was declared which virtually gave the government into the hands of the British. In 1853 the young Mahárája attained his majority, the administration was intrusted to him, and his territories have since been governed well.]

211. **Bhopál.**—Bhopál, 8200 square miles in extent, embraces a population composed chiefly of Hindus and Pathans of about 800,000. Its southern boundary is formed by the Narbada (Narmadá), from which the country rapidly rises northward to the Vindhya hills, and forms on the northern side a plateau sloping, as the direction of the rivers indicate, towards the north.

*Towns.*—**Bhopál**, the principal town, is the residence of the Bégam. **Sihor** is a military town, and the residence of the Political Agent.

[The state of Bhopál was founded by a Pathan adventurer during the reign of Aurangzeb. Throughout the Márátha wars it remained friendly to British interests, and suffered from the attacks of the Márátha troops. In 1818 it was taken under British protection. The present ruler is called the "Nawáb Bégam of Bhopál."]

212. **Rewa.**—Rewa or Baghelkhand (= the Country of the Baghels) lies between the North-Western and the Central Provinces. It extends over an area of little less than 18,000 square miles. The greater part of the

country is mountainous, and rises in three successive plateaus or vast terraces from the valley of the Ganges The first has an average elevation of 500 feet above the sea; the second of 1000; and the third is of much greater altitude. The first plateau, formed of sandstone, is of great sterility; the second is well cultivated, and produces crops of grain; and the third has an exceedingly fertile soil. There are several rivers flowing through the country, but none are navigable. The produce is chiefly wheat, barley, pulse, and cotton; coal and iron are found in considerable quantity. The population, which amounts to about three and a half millions, consists for the most part of Rájputs. Rewa is under the government of a chief styled a Mahárája.

*Towns.*—**Rewa**, the chief town, is but a poor place, with a scanty population.

213. **Minor States of Central India.**—DHÁR contains an area of about 1000 square miles, and a population of 100,000. Some portions lie very remote from others. It produces abundant crops of **rice, wheat, millet, oil-seeds, sugar-cane, tobacco, opium,** and **cotton.** Dhár is ruled by a chief. The principal town is **Dhár.** RATLAM is about 1000 square miles in extent, and is governed by a chief. **Ratlam** is the principal place. JHABBUA, JAURA, and some few other unimportant districts, are also included in Málwa.

214. **Bundelkhand.**—Bundelkhand (= the Country of the Bundelas) consists of the British districts of **Bánda, Jhánsi,** Jalaun, and Hamirpur, all under the government of the North-West Provinces, of several native states, and a number of petty jághirs. The native

states are all subject to the political superintendence of the Viceroy and Governor-General's Agent at Indor (Indaur). The country is traversed by several ridges of hills, and is well watered by a number of streams which flow into the Jamna.

*Towns.*—**Chatterpur**, a thriving town, has manufactures of paper and coarse cutlery. **Ajaigarh** (= Impregnable Fortress), a hill-fort, has near it some remarkable ruined Hindu temples.

---

## AJMÍR AND MHAIRWÁRA.

215. These two small tracts, together covering an area of 2700 square miles, are under the government of a Chief Commissioner appointed by the Government of India. The people, about 400,000 in number, are for the most part Hindus. The prevailing languages are Hindi and Urdu.

**Ajmír** is surrounded by the Rájput States of Jodhpur, Kishangarh, Udaipur, and British Mhairwára. The south-east part is level and sandy, but the northern and western portions are traversed by ridges of the Arávali mountains. The few streams which water the country are for the most part mere torrents in the rainy season, disappearing a short time after the rains have ceased; cultivation is dependent, therefore, on the tanks which are scattered in numbers over the district. The principal objects of culture are rice, wheat, barley, millet, and cotton. Iron and lead are found, the former in abundance; and the soil in many places is strongly impregnated with mineral salts, chiefly carbonate of soda.

**Mhairwára** is a narrow stretch of land lying between Udaipur and Jodhpur, and forming part of the Arávali mountains. The lower parts of the valleys between the ridges are elevated about 1600 feet above the sea, and the summits of the hills rise in some cases to 1000 feet more. Lead, copper, and antimony ores occur, and iron is found in inexhaustible quantities. Cotton, opium, rice, wheat, millet, and barley form the staple crops.

[The *Mhairs*, the inhabitants of this tract, were for ages a wild, lawless race, somewhat like the Bhíls in habits. Under British rule they have forsaken their lawless ways, and settled down to the peaceful arts. Education is making progress; Ajmír has a Government college, and there are numerous Government and missionary schools.]

*Towns.*—**Ajmír** is the chief town; **Násirábád** is next in importance.

---

## THE BOMBAY PRESIDENCY.

216. *Boundaries.*—This Presidency is bounded on the north by the Panjáb; on the east by the dominions of Sindia, Holkar, the Nizám, and the Presidency of Madras; on the south by the Madras Presidency; and on the west by the Arabian Sea.

217. *Dimensions.*—The Bombay Presidency is formed by a long and narrow strip of country of irregular breadth, stretching along the western shores of India,

and, including tributary states, embraces an area of 200,000 square miles.

218. *Climate.*—The climate varies much in different parts. In the districts along the coasts it is very sultry, the lowest temperature being 70° F., and the average annual heat 80°. In the same districts also an enormous quantity of rain falls.

[At Bombay the average rainfall is eighty inches; at Thána and Ratnágiri more than one hundred inches; and at Mahábaléshwar, where more rain falls than in any other part of Western India, as much as 248 inches have been known to fall in a single year. On the setting-in of the south-west monsoon, the rain comes down in torrents, often incessantly for days together, accompanied with lightning, thunder, and violent wind on the west side of the mountains, while on the east side, at only thirty miles' distance, perhaps not a drop of rain has fallen. The average rainfall in the Poona (Púna) collectorate does not reach twenty inches, and in the hot season this part of the country, from the absence of trees, presents a most desolate aspect, appearing as a parched-up waste. In Broach (Bhroch) and the northern collectorates the rains are moderate, not much exceeding thirty inches, but in Gujarát, still farther north, they are sometimes so scanty that all the horrors of famine ensue. In the same part of the Presidency the thermometer often falls to 40° during the cold weather, but from March to the end of May hot winds prevail, and it then frequently rises to more than 110°.

219. *Natural Productions.*—VEGETABLE PRODUCTS.—**Cotton** is grown almost throughout the Presidency, but especially in the southern Márátha country and Broach (Bhroch). **Rice, millet, barley,** and **gram** are the chief ordinary crops; the **sugar-cane** flourishes in several parts, and **coffee** in Belgaum (Bilgáon); **wheat** is cultivated in Gujarát, the **potato** in Poona (Púna) and Kaira; **indigo** in Khándesh, and **tobacco** and **oil-seeds** almost everywhere. The hills are covered

with forests which furnish many valuable kinds of timber, as also gums, drugs, and dyes, in great variety. Cocoanut-palms form a perfect fringe along the coast, and plantains, mangoes, and all the common Indian fruits, are everywhere grown in abundance.

ANIMALS.—The animals are such as are generally met with throughout India; **tigers** are common in several parts, and are very destructive; there are also **elephants, cheetahs, wolves, hyenas, jackals, wild boars, deer**, and **monkeys**; **snakes** of a venomous description abound; **birds** are numerous and of great variety; **fish** is caught in all the streams and in the sea; and amongst the domestic animals there are horses, camels, oxen, buffaloes, swine, goats, and sheep.

220. *Population.*—The population of the Presidency of Bombay and its subordinate states is estimated at **25,000,000.**

[Besides Máráthas and Gujarátis, who are very numerous, the population includes numbers of Bhíls and other rude tribes, Pársis and Europeans.]

221. *Religion.*—**Hinduism** is the most general religion, but **Muhammadans** are numerous, as also are **Pársis** and **Christians. Jains** are here more common than in any other part of India.

[The Bhíls and other aboriginal tribes occasionally profess either the Hindu or Muhammadan faith, but they seem to know little concerning them except the name.]

222. *Languages.*—**Máráthí** and **Kanarese** (Kannadam) are used chiefly in the centre and south respectively; **Gujaráthi** is the language most used by the

merchants in the north of Western India; **Sindhi**, a dialect of Sanskrit, is the language of Sindh, though west of the Indus the common speech is **Baluchikí**; **Portuguese, Arabic, Persian, Urdú**, and **English** are also employed.

223. *Education.*—Education is under a Director.

**[Bombay has a University, and there are Colleges at Bombay and Poona (Púna). The annual expenditure on public instruction is about twenty-one lakhs of rupees.]**

224. *Industry.*—Agriculture forms the occupation of the great bulk of the people. The only manufacture of importance is silk, which, widely held in repute as "Pitambur," is made chiefly in Thána, Násik, and Poona (Púna). Cotton-weaving is carried on in factories in the European style to some extent in the capital of the Presidency, and coarse pottery almost everywhere.

225. *Means of Communication.*—No part of British India perhaps is so deficient in good roads as this Presidency, though so much has been done of late to improve the means of transit. Except in a few cases, the passes over the mountains are impracticable for carts, and, in the plains, roads which are good in the dry season are often rendered impassable by a heavy fall of rain; in the wet season all traffic is stopped.

**[Railways.—The *Bombay, Baroda, and Central India Railway* extends from Bombay to Ahmadábád, and thence to Wadhwán, in Gujarát, a distance of 390 miles. The "*Great Indian Peninsula Railway*" extends from Bombay to Kalián, a distance of 33 miles, and then branches off north-east by the Thull Ghát to Bhusáwal, and is thence continued in the same direction to Jabalpur, and eastward to Nágpur; another branch from Kalián**

crosses the mountains by the Bhor Ghát, passes south-east to Poona (Púna), and terminates at Raichúr (Ráyachúrú), where it forms a junction with the railway from Madras. The *Dhond and Manmád Railway* forms a chord line 145 miles in length, and serves to connect the railway systems of Northern and Southern India without the necessity of ascending and descending the Western Gháts. The *Sindh Railway* extends from Karáchi to Kotri, a distance of 105 miles, where it is met by the *Indus Valley State Railway*, 500 miles in length, which carries the line along the right bank of the Indus to Sakkar, and thence along the left bank of the river till it reaches the terminus of the Panjáb Railway at Multán.]

226. *Commerce.*—IMPORTS.—**Cotton, woollen,** and **hardware goods**, and **machinery** from the United Kingdom; **shawls** from Kashmír; **opium** from Málwa; **silk** and **tea** from China. EXPORTS.—**Cotton, wool, Kashmír shawls,** and **seeds** to Great Britain; **cotton** and **opium** to China. The chief trade is with London and Hongkong; but considerable commerce is also carried on with the Persian Gulf, Calcutta, Malabár (Malaiyálam), Aden, and the opposite coast of Africa. PORTS.—The chief ports having foreign trade are Karáchi, Dholerá, Kambay (Khambáyat), Broach (Bhroch), Súrat, Bombay, Ratnágiri, Vingurla (Vengurlí), and Kárwár.

227. *Government.*—The government of the twenty-three districts into which the Presidency is divided is vested in the "**Governor of Bombay in Council,**" subject to the control of the Government of India. The various tributary states within the limits of the Presidency are governed by their own chiefs or princes, who are more or less subject to British rule.

228. *Revenue.*—The revenue is derived from land,

opium, salt, customs, stamps, and excise, and amounts to about **ten crores of rupees annually.**

229. *Condition of the People.*—In this Presidency the social condition of the great bulk of the population is much the same as that of Madras.

Peasant's Hut, Bombay Presidency.

230. *Divisions.*—Bombay is for administrative purposes divided into four great divisions—the *Northern*, the *Southern*, the *Kanarese*, and the *Sindh* Divisions. The **Northern Division** consists of the districts of Ahmadábád, Kaira, Panch Máháls, Súrat, Broach (Bhroch), Khándesh, Násik, Thána, and Kolába; the **Southern Division** of Ahmadnagar, Poona (Púna), Sholápur, Sáttára, and Ratnágiri; the **Kanarese Division** of Kaládgi, Dhárwár, Belgaum (Bilgáon), and North Kanara (Kannada); and the **Sindh Division** of Karáchi, Thar and Párkar, Haidarábád, Shikárpur, and Upper Sindh Frontier. Subject to the Government of Bombay are Khairpur, Kach, Kathiawár, and several

petty states in Gujarát, Kolhápur, Sáwant Wárí, and the Márátha Jághirs.*

231. **Bombay**, the capital of the Presidency, is situated at the south-east extremity of the island of the same name. The island is composed of two ranges of rock of unequal length, with an intervening valley about three miles wide, which has been converted from an unwholesome swamp into a salubrious residence. It is connected with Salsette by an artificial causeway. Good roads traverse the island in every direction, and many parts are studded with a pleasing variety of spacious houses standing amidst gardens and groves of fruit-trees. The scenery in the neighbourhood is charming. The population amounts to 640,000, and is of the most varied character, including Hindus, Muhammadans, Pársis, English, Portuguese, Chinese, Malays, and people from every part of Asia, speaking their native languages and wearing their national dress. In the European portion the houses are three and four storeys high, and there are many churches, mosques, and other fine buildings; in the native portion the houses are small and the streets narrow, dirty, and crowded. Bombay harbour is the largest and safest in India, and has every requisite for a great seaport. It is easy of access, affords excellent anchorage, is sheltered, capacious, and admirably adapted for docks. Ships of considerable size are here constructed. The commerce is

---

* Besides the states within the geographical limits of the Bombay Presidency, there is under the supervision and control of the Bombay Government the settlement of *Aden*, with an area of about eleven square miles, a population of 19,000 souls, and a yearly revenue of 60,000 rupees.

of great extent and importance, and its merchants are amongst the wealthiest in India. Bombay trades chiefly with Great Britain, China, Persia, the Arabian Gulf, Africa, France, and America. Steamers carrying mails and passengers ply regularly from Bombay to Suez (or to Europe), Point de Galle, the Persian Gulf, and all the coast ports round to Calcutta. Near Bombay are the remarkable cave temples of Keneri and Elephanta.

**[The island of Bombay came into the possession of the Portuguese in 1530, and was by them ceded to Charles II. of England in 1661 as part of the dowry of his Queen (Catharine). In 1668 it was transferred to the East India Company "in free and common soccage, as the manor of East Greenwich, on the payment of the annual rental of £10 in gold, on the 30th September of each year." Though repeatedly besieged during the long struggles for supremacy in India by the Muhammadans, the Portuguese, and the Maráthas, Bombay has, from the time of its cession, steadily remained in the hands of the British.]**

232. **Ahmadábád.**—The hilly tracts of the Mewár province terminate on the extreme north of this collectorate; from thence the country undulates southwards, subdividing into flat open plains terminated by a range of hills about 700 feet high. The soil is very fertile, and produces wheat, cotton, millet, gram, sugar-cane, and rice. The water for irrigation is drawn chiefly from wells. There are no canals; the roads are poor owing to the want of proper material; and the only considerable river, the Sábarmati, is unnavigable.

*Towns.*—**Ahmadábád,** the principal place in the district, is situated on the Sábarmati. It was founded in A.D. 1412 by Ahmad Sháh of Gujarát, and at the com mencement of the 17th century was one of the finest cities in India. Numerous decayed mosques, palaces, mausolea, aqueducts, fountains, and serais still attest its former grandeur; and the Juma Masjid or great mosque, the mosque of Sujat Khán, and the "ivory" mosque are amongst the most magnificent structures in the land. The prosperity of Ahmadábád was swept away amidst the Márátha struggles for supremacy consequent on the decay of the Mughal empire. On the defeat of the Péshwa in 1818 it came into the hands of the British, and latterly has recovered somewhat of its former importance. **Dholerá** is a port for the shipment of cotton. **Gogo**, on the coast, has a harbour affording excellent anchorage. Near it is the island of Perim, noted for the large number of fossil bones found there.

233. **Kaira and the Panch Máháls.**—These districts much resemble in aspect that of Ahmadábád. They are for the most part perfectly flat, and destitute of canals, navigable rivers, and good roads. They produce, besides the usual grains, tobacco, indigo, cotton, and opium.

*Towns.*—**Kaira**, formerly the principal town, is of considerable size, situated in the midst of a fertile and beautiful district. **Nariad** is now the chief town.

234. **Broach** (Bhroch) and **Súrat.**—These two small collectorates closely resemble each other in every respect. Cotton, jowári, wheat, rice, and tobacco are grown. Leather, coarse paper, and cotton cloths are manufactured to a trifling extent. The roads are of the same sandy character as those of Kaira, but owing

to the flatness of the country communication between the different parts, except in the rainy season, is uninterrupted.

*Towns.*—**Broach** (Bhroch), the chief town in the collectorate, is situated on the Narbada (Narmadá), thirty miles from its mouth. It is a large place, and exports great quantities of cotton. **Súrat**, on the Taptí, is the principal place in the district. Formerly it was of great commercial importance, and merchants from all parts of Europe and Asia met there to traffic. Gradually its trade has been diverted to Bombay, but it still exports much cotton and grain. The English built a factory here in 1612.

[The districts of Ahmadábád, Kaira, Broach (Bhroch), and Súrat, together with Kathiawár and Baroda (Barodrá), form the old province of Gujarát.]

235. **Khándesh.**—Khándesh may be regarded as a great basin formed by the Sátpura range on the north, elevated ground on the east towards the Nizám's dominions, the Sátmála and Ajanta ranges on the south, and the Sáhyádri Mountains or Western Gháts on the west. Many streams rise in the high grounds on the borders, and flow into the Taptí, which traverses the district from east to west. The soil of the low grounds is fertile, and produces all the ordinary crops, but much of the cultivable portion is overrun with jungle. The villages are few and widely scattered.

[Besides the Maráthas, who form the bulk of the population of Khándesh, there are great numbers of Bhíls, a rude race occupying the hills and jungles along the Narbada (Narmadá) and the Taptí, and the hilly portions of Gujarát and Málwa. They are considered the remnant of the aborigines who were driven from the plains by the Bráhmanical Hindus. They are very black, short, and slender, but hardy and agile, and go about constantly armed with knives and bows and arrows. Their clothing consists of a narrow coarse cloth wrapped round the loins, and their dwellings are huts of the rudest description, constructed of sticks and boughs fastened together with long grass. A few such huts huddled together and surrounded by a small patch of cultivation form a Bhíl village. Until of late years, the Bhíls were marauders and murderers, and lived by rapine. Their practice was to rush suddenly down from their mountain fastnesses, and, sweeping over the open country, committing the most frightful barbarities upon the peaceful inhabitants, to carry off to their retreats, beyond the reach of reprisal, cattle and everything movable they could find. Not the least of the blessings conferred by the British Government on this part of India is the security from all such perils now. The Bhíls, though still very uncivilised, have been reduced to order; the more turbulent have become soldiers, and the remainder subsist by agriculture and the exercise of the peaceful arts. The Bhíl country was anciently called *Nishadha.* Nala, the husband of the beautiful Damayanti, was once rája of this tract.]

*Towns.*—**Chopra** and **Dhulia** are the largest towns in the district.

236. **Násik.**—The Gháts form part of the western boundary of Násik, and send out numerous spurs forming beautiful valleys intersected with streams flowing to the east, in which direction the country gradually sinks into plains.

*Towns.*— **Násik,*** the chief town, contains many

* This place, anciently called *Panchavati*, was for some time the residence of Ráma during his exile. According to the "Rámáyana," the district was then occupied by a number of Rákshasas detached by Rávana, king of Lanka, to guard his frontiers, under the orders of his brothers Khara and Dúshana. The latter were accompanied by their sister Súrpanakhá, who one day meeting with Ráma, so much annoyed him by reviling his wife Sítá, that in a moment of angry thoughtlessness he cut off the Rákshasa's nose, hence the modern name Násik—*the mutilated.* The war with Rávana arose out of this circumstance.

ancient Hindu temples, and is regarded by Bráhmans as the peculiar seat of piety. From its position with respect to the rest of British India, from its accessibility, salubrity, and security from sudden foreign attack, it has been pointed out as the fittest position for the seat of the Government of India.

237. **Thána** and **Kolába.**—These two districts lie along the coast, and have the Ghâts for their eastern boundary. Their respective chief towns are **Thána** and **Alibhág.**

238. **Ahmadnagar.**—This district lies south-east of Násik, between Poona (Púna) and the Nizam's Dominions.

*Towns.*—**Ahmadnagar** is a town of some importance. It was captured in 1803 by General Wellesley after a two days' siege.

239. **Poona** (Púna).—This collectorate forms an elevated tableland, interspersed with isolated hills, and crossed by several offshoots from the Western Ghâts, which take a general south-east direction, and sink gradually into the plains of the Nizam's Dominions. The district is traversed by some good roads; is well watered by the Bhíma (*Bhíma* = the Terrible) and its tributaries; and produces rice, maize, millet, cotton, sugar-cane, and potatoes. The manufactures are of little importance; paper is made at Poona (Púna), and some coarse cotton fabrics are produced in the larger towns.

*Towns.*—**Poona** (Púna), the chief town, is situated in a treeless plain, 2000 feet above the sea. It contains the former residence of the Péshwas, whose capital the city was, a college, and many public buildings. It is an important military post, and its cantonment one of the largest and best arranged in India.

**[Khándesh, Násik, Ahmadnagar, Poona (Púna), and Sholápur formed part of the Márátha dominions, and fell to the British in 1818, on the renunciation of all sovereign power by Báji Rao, the last of the Péshwas.]**

240. **Sholápur.**—Sholápur presents a succession of rising-grounds and valleys watered by the Krishna and the Bhíma. Cotton forms its staple product, and is sent in enormous quantities to Bombay.

*Towns.*—**Sholápur,** the chief town, is situated in the east of the collectorate. **Bársi** is a depot for the cotton passing from the western districts of the Nizám's Dominions to the Great Indian Peninsula Railway. **Pandhápur** has a temple of peculiar sanctity, dedicated to Vishnu, to which great crowds of pilgrims annually resort. Here in 1815 Gangádhar Sástri was murdered at the instigation of Trimbakji Dainglia.

241. **Sátára.**—Sátára is divided into two parts by a chain of hills branching off from the Mahádeo range, and running parallel to the Western Gháts nearly to the banks of the Krishna. The district to the west is hilly, well cultivated, and productive; that to the east, on the contrary, consists of wide open plains, yielding little more than pasture. The western part is well watered by a number of streams which fall into the Krishna, and by the copious rains brought by the south-west monsoon; the eastern portion is very subject to drought. The productions are cereals, coffee, flax, cotton, and sugar.

[Portions of Sátára, varying from 400 to 1000 square miles in area, are possessed by Jághirdárs, or chiefs who draw the revenues, perform military service, and in many other respects resemble the feudal barons of Europe during the Middle Ages.]

*Towns.*—**Sátára**, the chief town, stands in the midst of a rugged country, in a deep hollow nearly surrounded by hills.

[Soon after Báláji Báji Ráo became Péshwa, he usurped the supreme authority, and kept the rája and his family in strict captivity. In this humiliating state the descendants of Siváji remained for many years, until at length the army of the Péshwa was surprised and defeated at Ashti in 1818 by a British force, and the rája and his family fell into the hands of the victors. They were at once set at liberty, and next year, on the close of the war, the British Government raised the fallen dynasty by assigning Sátára to the Rája Pratáb Singh. He bound himself to hold his territory subordinate to the British power, not to increase or diminish his military force without its sanction, and to abstain from holding all intercourse with persons not his subjects, except through the Resident. The British charged themselves with the defence of his kingdom. The rája violated his treaty, was deposed in 1839, and died at Benares (Banáras) in 1847. His brother Shâji succeeded, and died in 1848 without leaving issue. Sátára was then annexed to the British dominions.]

242. **Kaládgi, Dhárwár**, and **Belgaum** (Bilgáon).—Much the greater part of these collectorates consists of extensive plains of black soil, well adapted for the growth of cotton. Cotton is, in fact, very largely cultivated, and forms the principal product.

*Towns.*—**Bíjápur**, in Kaládgi, a large decayed city, is an immense mass of ruined tombs, palaces, mosques, and other buildings. It was once the capital of a powerful kingdom, and contained, it is said, nearly a million inhabited houses and 1600 mosques. It is now comparatively abandoned, the population numbering about 13,000 only, and for the most part very poor. **Belgaum** (Bilgáon) and **Dhárwár** are the chief towns in their respective collectorates. **Húblí** is a place of considerable trade, and the cotton mart for the Northern Parganas, as **Bankápur** is that for those of the south.

243. **Ratnágiri.**—This collectorate, called also the Southern Konkan, forms a long and narrow strip of country, sloping upwards abruptly towards the west into hills varying from 2000 to 4000 feet above the sea. The hilly tracts are exceedingly rugged, being intersected by ravines covered with jungle infested with wild beasts. The passes generally are impracticable for vehicles, and are traversed only by pack-bullocks. There is no river of any importance, but numerous streams rising in the hills find their way to the coast, and break it up into numberless small shallow harbours, in which, previous to British rule, lurked those pirates who were a terror to the merchants on the western coast. Much of the surface is rocky and sterile, and along the banks of the rivers only is fertile soil to be found. In these places rice is grown, the inferior grains being raised on the higher grounds, which are poor, and yield such scanty crops that the cultivation is shifted from place to place every year. The chief towns are **Ratnágiri** and **Vingurla** (Vengurlí), both upon the coast.

[Dháwar, Belgaum (Bilgáon), and the Konkan came under British rule in 1818.]

244. **North Canara** (Kannada).—This collectorate, the most southern portion of the Bombay Presidency, is one of the most fertile tracts in India. The plains

are studded with cocoanut-palms and rice-fields; the hill-slopes produce cardamoms, pepper, and areca-nuts; and the summits of the Gháts are crowned with dense forests of teak and other valuable woods. In consequence of the profusion of vegetation the hilly districts are very unhealthy, and the population therefore exceedingly sparse. The greater part of the people live in the plains and along the coast, but in the whole province the average number of inhabitants to the square mile does not exceed eleven.

**[Many small rivers run from the mountains to the sea, and two of them in their descent form magnificent falls. Some distance above the decayed town of Gásopa, the Sherávati, which enters the sea at Honáwar, divides into several channels, and forms four falls, all of them very beautiful and distinct in character. One, called "The Grand Fall," passes over a ledge of rocks, and falls perpendicularly from a height of 880 feet. On the Taddry river, which enters the sea a few miles to the north of Kumpta, are "The Lushington Falls."**

*Towns.*—The towns in North Canara (Kannada) are small and few, the agricultural population preferring to live on their farms. The principal places are situated on the coast, and are inhabited chiefly by Mápillas and Konkani Bráhmans, who trade and keep shops. **Kárwár**, on the coast, is admirably adapted for a harbour. **Kumpta** is a port for the shipment of cotton brought from Bellary (Ballári) and the Southern Maráatha country.

## SINDH.

245. *Physical Features.*—**Sindh**, which is familiarly called "Little Egypt," from the physical resemblance between the two countries, comprises the lower course and delta of the Indus, and covers an area of 52,000 square miles. Its great feature is of course its river. On the melting of the mountain snows this great stream rushes furiously down from the magnificent range of the distant Himálayas to the sea, breaking down banks, and whirling along with it trees and every other object coming within its influence. In the winter months it is calm and sluggish, and its water is drawn off by innumerable canals and watercourses to the neighbouring lands, which are not exceeded in fertility by any tract of country anywhere known. The delta of the Indus lies low, and at every tide the sea-water rushes for miles up the numerous creeks and smaller branches of the river, rendering the fresh water brackish, and the soil so salt as to be unfit for agricultural purposes during the greater part of the year. East of the Indus, and beyond its fertilising influence, the country becomes a desert of hard clay, interspersed with sandhills covered with coarse, nutritious grass, affording food for herds of camels, oxen, and buffaloes, and flocks of sheep. West of the Indus a few limestone and salt ranges are the only heights met with till the valley terminates in the mountains of Baluchistán.

246. *Climate.*—The climate is remarkably dry and sultry.

[At Sakkar the summer temperature is 102° F., and even the

waters of the Indus attain a warmth of 92°, while farther north the heat is so great that the Afgháns say "the sun of Sindh will turn a white man black and roast an egg." The hot season lasts from March to September; the cold from October to March. During the latter, in the northern districts, frosts occur. Rain seldom falls, the country being beyond the influence of both monsoons. One consequence of this is an incredible quantity of dust; the slightest wind raises it in clouds; and frequently dust storms occur that almost defy description.]

247. *Productions.*—The arable parts of Sindh produce two crops a year; that which is sown in spring and reaped in autumn consists of those products which require considerable heat to bring them to maturity, such as **rice, maize, cotton, sugar,** and **indigo**; and that which forms the spring harvest consists of **wheat, barley, millet, oil-seeds, hemp,** and **tobacco**. In the districts not affected by the inundation, the crops are artificially irrigated, in some places by simply opening canals and drains, in others by means of wheels. Sindh is rich in animal life. **Tigers, hyenas, jackals, deer** of all kinds, wild asses, and wild hogs, are common. Vast herds of buffaloes are bred in the swampy tracts, and, strangely enough, great numbers of **camels** also. **Horses,** small and poor-looking, but very enduring, and sheep and goats, are also reared in great plenty. **Birds** are in great variety; **alligators** are met with in the Indus; **fish** forms the chief food of many of the people; snakes are common, and every pool swarms with leeches.

248. *Manufactures, Trade.*—The manufactures are important; they consist of silk, cotton, and mixed cloths, paper, leather of a very solid and durable quality, swords, and firearms. Earthenware is made

in all the towns. The natives excel as weavers, turners, and dyers, and their wooden lacquer-work is known throughout India. The trade of the country is rapidly increasing. The *imports* consist of **piece-goods, silks**, and **woollens**, chiefly from Bombay; **cotton** and **silk fabrics** from the neighbouring countries; **raw silk, drugs, dyes, gold thread**, and **horses** from Khorásan, Kábul, and Bokhára; and **ivory** and **wood** from the Malabar coast. Its *exports* comprise **rice** and other grains, **indigo, assafœtida**, and **hides** by sea; **silk, cotton**, and other fabrics, **indigo, metals, sugar**, and **spices** to Khorásan and the Panjáb.

*Towns.*—**Haidarábád**, the capital, is an extensive place, with a large brick-built fort, standing on a stony eminence near the Indus. In its streets may be seen natives of every country in the north-west of India. Small lacquered betel and snuff boxes of pretty and curious workmanship are made here in very great quantity. Six miles north of Haidarábád is the village of **Miáni**, the scene of the battle fought in February 1843 between the Amírs and the British, in which the forces of the latter, consisting of less than 3000 men, led by Sir Charles Napier, gained a complete victory over the former at the head of 22,000. **Karáchi**, the chief port, is situated on the western extremity of the coast. Its position is one of such great commercial, political, and military consequence as to make it, as it were, the gate of Central Asia. It is sometimes called the Liverpool of India. A railway connects it with Haidarábád. **Táttá** (Tháthá) stands between Karáchi and Haidarábád. It was in olden times wealthy and important, a fact proved by the numbers of splendid mausolea that still remain. It is now a miserable and unhealthy place, inhabited chiefly by beggars. **Lárkhána** is one of the principal grain marts of the country; it has some silk and cotton manufactures. **Rohrí** has many mosques, in the largest of which is kept, in a richly jewelled case, a single hair in amber; the pious Musalmán believes it to be a hair of the beard of Muhammad. Opposite to Rohrí stands **Sakkar.** **Shikárpur** is one of the most important commercial towns in Sindh. Its inhabitants are chiefly Hindu merchants, who have commercial connections all over the East. It is at the junction of the routes leading to the Panjáb on the north, to Karáchi on the south, to Márwár on the east, and to Kandahár on the west. **Amarkot** (*Amara kota* = Fort of the Immortals) is noted as being the birthplace of the Emperor Akbar.

[The history of Sindh is little more than a record of invasion. Time after time this unfortunate country fell under the despotic sway of foreign conquerors, till in 1779 the Talpur tribe of Baluchis rebelled against the then reigning Nawáb, and wrested the government from his hands. The Talpur chiefs or Amírs parcelled out the country amongst themselves, and governed it so oppressively that, more than ever, fear and misery fell upon the people and impoverishment upon the land. Several commercial treaties were entered into at different times between the British and the Amírs, but by the latter always with reluctance. Gradually their coldness and jealousy grew into hatred, and in February 1843 their hostile wishes turned to hostile acts. British officers were openly insulted and the Resident was attacked. After a few decisive actions, Sindh was completely subjugated and annexed to the British Dominions.]

249. **Khairpur** is 6000 square miles in extent, and contains a population of 130,000. Geographically it

forms a portion of Sindh, and is like it in physical aspect.

*Towns.*—**Khairpur** is little more than a large collection of mud hovels; it contains the palace of the "Mír," by whom the territory is governed.

---

## KACH.

250. *Physical Features.*—Kach comprises an area of 6500 square miles. It is a long narrow peninsula between the Gulf of Kach and the Ran. Through the middle, from east to west, stretches a range of hills called Lankhi, and more to the north and parallel to the first runs another similar range. The hills are volcanic, and seem only recently to have cooled from a state of fusion; the masses of rock are bare and black, and thrown together in wild confusion, in many parts presenting wide chasms, down which, during the rains, rapid torrents sweep into the lower grounds.

251. *Climate.*—The climate is healthy, temperate, and agreeable for nine months in the year, but in the hot weather the temperature is high; in the winter the cold is frequently great, and water if exposed at night in open vessels turns to ice.

252. *Productions.*—As in Sindh, water is scarce, and except in the rainy season all vegetation fails, save here and there patches of stunted brushwood and brambles. A few fields in the vicinity of villages are cultivated, and a tract of land along the north side of

the province produces most luxuriant pasturage; the rest of the country during the greater part of the year presents nothing but a rocky and sandy waste. The produce of the land under cultivation is not sufficient to support the scanty population, so that Kach even in the best seasons is dependent upon Sindh for supplies of grain. Rice, millet, sugar-cane, cotton, together with grapes and musk-melons, which are famous, are the chief vegetable products. This country is famous for a singular breed of horses, and the wild ass, a beautiful creature, is found in herds in the Ran.

253. *People, Religion.*—The inhabitants exceed half a million in number, and are partly Hindus and partly Muhammadans.

254. *Government.*—The Government of Kach is held by the "Ráo;" under him are a number of Jádeja chiefs, who hold their lands on military tenure, and exercise authority within their respective domains; they form what is called the bhayád or brotherhood of the Ráo, and advise him in political matters.

255. *Towns.*—**Bhuj** is the capital. In June 1819 both the city and the fort, about a mile distant, were nearly destroyed by an earthquake. **Mándvi**, the principal seaport, is on the south coast; it has considerable trade.

[The *Ran*—a corruption of the Sanskrit *aranya*, a waste—which separates Kach from Kathiawár, is one of the most singular geographical features of India. It covers a surface of 7000 square miles, and is neither a swamp, a desert, nor a lake, but a combination of all three; in the dry season it presents the appearance of a sandy waste, interspersed with wide sheets of salt, shallow pools of water, ridges of sand, and patches of vegetation; in the rainy season it is covered with water knee-deep. Several roads, passable by vehicles, cross it in the narrowest portions, but except in the rainy season, to traverse it in the daytime is almost certain death. Wild asses are here found in great numbers, as also are apes, porcupines, and vast flocks of birds. Myriads of musquitoes, flies, and other insects torment the traveller at all times, but on the subsidence of the water, when multitudes of dead prawns and fish remain strewed over the surface, they become so intolerable that no horse will face them.]

---

## GUJARÁT.

256. **Gujarát.**—Gujarát (the ancient *Sauráshtra* = the Land of the Sauras) comprehends **Baroda** or the Gaikawár's territory, Kathiawár, Máhi Kánta, Rewa Kánta, and many petty states, covering a total area of 50,000 square miles. It is for the most part flat, and produces cotton, rice, wheat, barley, millet, gram, sugar-cane, and fruits in abundance. Amongst wild animals there are tigers, leopards, wolves, and hyenas; deer are very numerous, and waterfowl in great variety. Camels, buffaloes, oxen, and horses are common; the wild ass is found in the uncultivated tracts, as are wild cattle resembling the bison. The country is watered by the Sábarmati, Máhi, Narbada (Narmadá), and Taptí. There are scarcely any made roads, and in the rainy season all traffic ceases. People of many tribes and races make up the population, which exceeds six millions; the ruling race is Márátha, but it is not the most numerous; Rájputs, Jains, and Bráhmans abound, the latter being mostly landed proprietors; Musalmáns and Pársis resort to the towns; and in-

habiting various parts are the Kulis, the Kumbis, the Kathis, and the Bhíls.

*Towns.*—**Baroda** (Barodrá), the capital, is a large trading city. **Kamba** (Kambáyat), at the head of the gulf to which it gives its name, is an ancient place now much decayed. **Dísa** is a British cantonment. **Patan-Somnát**, on the south coast of Kathiawár, contains the remains of a celebrated temple dedicated to Somnáth (= the Lord of the Moon). It was plundered of immense wealth by Mahmud of Ghazni, A.D. 1025. Near **Junágarh** is the sacred mount Girnár, a granite peak rising 2500 feet above the plain. On the summit are several Jain and Hindu temples. Great numbers of pilgrims and devotees ascend the mount daily, and in February a religious ceremony takes place, when thousands of people attend.

[When the Márátha power was at its height, a chief of that nation, called the Gaikawár, was appointed ruler of Baroda (Barodrá). His immediate descendants gradually extended their authority until under Dámaji Gaikawár it embraced all Gujarát. This prince, in attempting to throw off his allegiance to the Péshwa, was treacherously made prisoner, and compelled to purchase his liberty by ceding half his possessions. Subsequently hostilities occurred between the Péshwa and the Gaikawár, and the latter sought British assistance. An alliance was formed, and later the murder of the Gaikawár's minister at Pandhápur led to events which cost the Péshwa all his possessions and left him a captive, while his rival's dominions were preserved to him. The tribute of the whole of the petty states into which Gujarát is divided is collected by the British, who pay the Gaikawár his share.]

257. **Kolhápur** is a rugged tract of country, 3200 square miles in extent, sloping from the Western Gháts to the plains on the east. It is traversed by numerous streams and torrents which find their way

to the Krishna. The people are chiefly Máráthas. Kolhápur is governed by a rája subject to British authority. The population is 800,000.

258. **Sáwant Wári.**—This state, 900 square miles in extent, is of a rugged character, like the rest of the Konkan,—hills, rocks, forests, and rivulets presenting every variety of scenery. The products are rice, wheat, grain, and cocoa-nuts, which last are largely exported. Several good roads lead to the coast. The territory is governed by a Már´atha chief called the "Sir Desai," whose annual revenue amounts to about three lakhs of rupees.

---

## PORTUGUESE POSSESSIONS IN INDIA.

259. The Portuguese possessions in India are **Goa**, **Damán**, and **Diu**—all within the limits of the Bombay Presidency. The Goa district has an area of 200 square miles. It is well watered and fertile, and in physical characteristics resembles the neighbouring territories. The products are rice, pepper, cocoa-nuts, betel, and salt. The people, who number about 350,000, are chiefly the descendants of Portuguese. **Goa**, the once splendid capital of the wide-spread Portuguese dominions in Asia, is now a place of little consequence. The harbour, which has a magnificent appearance, is scarcely inferior to that of Bombay. **Damán** is a seaport town and district in the collectorate of Súrat. It has docks, and some shipbuilding is

carried on. **Diu** is now fallen to decay. It stands at the south-east end of the small island of the same name in Kathiawár.

---

## RÁJPUTÁNA.

260. This immense tract of country, extending from the deserts of Sindh on the west to Agra on the east, skirting the Bombay Presidency on the south, and stretching towards the Panjáb on the north, embraces an area of 130,000 square miles and 10,000,000 inhabitants. Excluding the small British territory of Ajmír and Mhairwára, situated almost in the centre of Rájputána, the whole of the Rájputána territory is divided into eighteen independent states.* All are under the political superintendence of an agent to the Governor-General, to whom each state sends a vakeel, and they are all ruled by Rájput princes, except Tonk, which is under a Musalman chief, and Bhurtpore (Bharatpur), which is a Ját principality.

261. *Physical Features.*—Though so large a country, Rájputána is one of the least important parts of India. Much of it is an utter waste of moving sand, destitute of all but occasional traces of vegetation and water, and therefore of inhabitants. The greater part of all the

---

* These are Bikánír, Jaisalmír, Márwár or Jodhpur, Alwar, Bhurtpore (Bharatpur), Dholpur, Karauli, Jaipur, Kishangarh, Tonk, Bundi, Kotah, Jhaláwa, Partábgarh, Banswára, Dungarpur, Udaipur, and Serohí.

western states is desert, and Bikánír is the poorest of all; but some tracts in the east are, on the contrary, fertile and highly cultivated, producing much corn, cotton, sugar, tobacco, and opium; and where pasturage is abundant, herds of camels and horses and flocks of sheep are reared. Rivers are few, the chief streams being the Luní, the Banas, and the Chambal; water is therefore mostly obtained from wells, which in the eastern parts are of little depth, but in the more arid parts of the west are sometimes two or three hundred feet deep. Salt lakes and brine springs are common in the west, and from them much salt is made and exported. The manufactures include cotton and woollen cloths and working in metals and ivory.

262. *People, Religion.*—The people are chiefly **Rájputs** (= Sons of Kings), a tall, vigorous, and athletic race with very martial habits. They claim to be lineally descended from the warrior caste of Kshatriyás, the descendants of the sun and moon, and many wear round their necks metallic images of the sun and moon on horseback. The feudal system prevails amongst them, and the people are much attached to their chiefs. In the western states the **Bhattis**, a tribe of Rájputs, are numerous. They are a dissipated race, much addicted to opium.

[The *Bháts* form another class; they are met with in all Rájput communities as bards, singing the praises of the families to which they are attached, and satirical songs concerning their rivals. The *Cháráns*, another class of the same people, have unbounded influence among their countrymen, to whom they act as priests and historians. The common belief is that total ruin befalls the person who in any way causes the least drop of the blood of a

Cháran to be shed; hence their influence. One of this class not unfrequently accompanies travellers to protect them from robbers, and should any such appear and disregard his warnings, he does not scruple to wound himself and throw his blood towards them, uttering at the same time the most frightful curses. In extreme cases Chárans have been known to put themselves to death. *Játs* are also numerous in Rájputána, and the remainder of the people are Bráhmans and Jains.]

*Towns.*—**Abu**, in Serohí, is a sanitarium located on Mount Abu, in the Arávali range, 5000 feet above the sea. The mount is much resorted to by pilgrims, chiefly Jains, and near the summit there is a superb group of marble temples, the principal one having cost **eighteen crores of rupees** in building. **Bikánír** is situated in the midst of a desolate country, and from a distance bears a beautiful aspect. It has some tall houses and temples, and a citadel having a number of towers, but most of the houses are mere mud hovels painted red. **Jodhpur**, the chief town of Márwár, contains many handsome edifices of stone. **Mecheri** is the principal place in Alwar. **Bhurtpore** (Bharatpur = Bharata's Town), a large town with narrow, crowded streets, is interesting as having been the scene of two famous sieges in 1805 and 1826, in the latter of which it fell. The strong fortifications which once surrounded it have been removed. **Dholpur, Karauli, Jaipur, Kishangarh,** and **Tonk** are the chief towns in their respective states. Jaipur (= the City of Victory) is one of the best built cities in India; the houses are finely constructed of stone, and the streets run at right angles to each other. **Bundi** is also a finely built town in the midst of a beautiful country. **Kotah** is on the Chambal. **Partabagh, Banswára, Dungarpur, Udai-**

pur, and **Serohí** are the principal towns in their respective states, but are otherwise of little importance.

View, Jaipur.

# THE PANJÁB.

263. *Position and Dimensions.*—The Panjáb extends from the river Jamna on the east to the Sulaimán Mountains on the west, and from Kashmír on the north to Rájputána and Sindh on the south. The total area is estimated at 102,000 square miles.

264. *Physical Features.* In physical aspect the Panjáb varies considerably. The north-east corner of the country is a maze of lofty mountains, undulating hills, and fertile valleys; the north-west angle is also hilly; the eastern parts are flat, with occasional rocky hills, and all the rest of the country, except the Deraját, which slopes down to the Indus from the Sulaimán Mountains, forms an almost uninterrupted plain, inclining imperceptibly south-west, from about 1600 feet above the sea at the town of Jhílam to 230 feet at Mithankot. This plain is divided by the well-known rivers which flow through the land into five great natural sections called **Duábs.** They are the **Sindh Sagar Duáb**, between the Indus on the west and the Jhílam, the Chenáb, and the Panjnad on the east; the **Jetch Duáb**, between the Jhílam and the Chenáb; the **Rachna Duáb**, between the Chenáb and the Rávi; the **Bári Duáb**, between the Rávi and the Trimba on the west and the Bias * and Gharra on the east; and the **Jalinder Duáb**, between the Bias and the Satlej. Of

* The Chenáb is the ancient *Chandrabhága.* The Bias is the Sanskrit *Vipása* and the Greek *Hyphasis;* while the Satlej is the ancient *Sata-dru* (from the Sans. *sata* = a hundred, and *dru* = a tree or branch)—*i.e.*, flowing in a hundred branches.

these, the Sindh Sagar Duáb is the most extensive, but the Bári Duáb is by far the most populous and important.

[Wherever water is available for irrigation, the soil is remarkably productive; elsewhere the country either wears a miserably arid appearance or is covered with low brushwood, jungle, and reed grass. The soil consists chiefly of sand and clay, and in some parts no stone so large as a pebble can be seen for many miles. On this account, and because of the flatness of the country, the rivers constantly wear away their banks and flow in other directions, never pursuing exactly the same course two years in succession. Towns and villages, therefore, except along the Indus, where limestone and other rocks occur, are usually located at some distance from large rivers, and cultivation is rarely or never carried on close to the banks.]

265. *Climate.*—The climate is characterised by long periods of drought. The rainfall is not great anywhere, but greatest at the base of the hills, and least nearest Sindh, where the soil, if productive, is rendered so by irrigation. The rivers are mostly fed by the melting of the Himálayan snows, and in July, when the rains are added to the then swollen streams, they inundate much of the country.

[The hot weather begins about the middle of April, and the heat is almost intolerable; greater, in fact, than in any other part of India, the thermometer indicating more than 100° F. Frequent dust-storms then occur; and on calm days spiral columns of dust arise and whirl onward for one or two miles before subsiding. In September the weather begins to moderate; in October it is temperate and agreeable; and from November to the middle of April it is cold. Frosts then occur at night, but during the day the thermometer rises to above 70°.]

266. *Productions.*—MINERAL PRODUCTS.—**Gold** is found in the sands of the Chenáb and the Indus; **iron** and plumbago of an inferior kind are obtained

in the hills; **nitre** is got in abundance from the alluvial plains; and **salt** of an impure kind from many localities. The Salt Mountains, a range between the Jhílam and the Indus, yield this last in great quantity, besides **alum**, **antimony**, and **sulphur.** VEGETABLE PRODUCTS.—The most important objects of culture are **wheat**, **barley**, **millet**, **rice**, **cotton**, **hemp**, **indigo**, **tobacco**, **sugar-cane**, and **pulses**. **Oil-seeds**, **melons**, **cucumbers**, and **saffron** are also largely grown; and fruits, such as **dates**, **figs**, **oranges**, **mangoes**, and others, are common. **Flax** thrives, and **tea** is grown on the hills. Large trees are scarce. ANIMALS.—**Tigers**, **leopards**, **hyenas**, **wolves**, **bears**, **jackals**, **foxes**, **deer** of all kinds, and **monkeys** are amongst the largest kinds of wild animal life. The chief domestic animals are **camels**, **buffaloes**, **horses**, and **sheep**. Birds of all kinds, including eagles, kites, pheasants, partridges, and waterfowl, are to be met with in abundance. Alligators, porpoises, and fish swarm in the rivers. Many venomous reptiles are common. The **silkworm** thrives well, and **bees** are kept in great numbers for the sake of their honey and wax, which are both of the finest quality. Insects of all sorts are very numerous; and a very minute creature, the sandfly, which makes its appearance in the hot season, is a much greater plague to mankind than the musquito.

267. *People.*—The population, estimated at about **18,000,000**, consists of **Sikhs**, **Játs**, **Gujars**, **Rájputs**, and **Patans**. The Sikhs are most numerous in the Bari Duáb; the Játs are chiefly agriculturists, and

are found in all parts of the country east of the Indus; and the Patans west of that river. The Gujars are devoted to agriculture.

268. *Religion.*—More than half the people are **Muhammadans**; the majority of the remainder are **Hindus**, and the rest profess the **Sikh** religion. The Sikhs have no distinctions of caste.

Head of aged Sikh.

269. *Language.*—The language is called **Panjábi**, and is based on Sanskrit. In the large towns it is a dialect of Hindustáni; in the villages it differs considerably; and in the south, near Sindh, it is much mixed with Sindhí. **Urdú, Persian,** and **English** are taught in the schools.

270. *Education.*—As in other parts of India, popular education is under the control of a Director. There are two Colleges, one at Láhor and another at Dehli. The Colleges and some of the High Schools prepare their pupils for the Calcutta University Examinations.

[More has been done in the Panjáb for female education than in any other part of India. Hindus, Sikhs, and Muhammadans alike send their girls to school, where they are taught by native female teachers. The total expenditure on state education in the Panjáb amounts to about ten and a half lakhs of rupees.]

271. *Industry.*—**Agriculture** here, as in most other parts of India, occupies the greater portion of the population, but the manufacturing industry is very considerable and important. **Silk** and **cotton goods** are extensively made in most of the large towns, especially in Láhor, Amritsar, Multán, Shujábád, and Leiah. The silks of Multán are noted in Indian markets. Cotton cloths of a stronger and more durable texture than those of Britain are made at Rahon and Hoshiápur. **Carpets** like those of Persia, shawls, little inferior to those of Kashmír, brocades, and other rich silks, and **arms** of excellent quality are made at Láhor.

272. *Internal Communication.*—The Panjáb possesses great facilities of transit in its many navigable rivers; and of late years the British Government have constructed hundreds of miles of first-class roads connecting all the large towns. The roads are replete with every convenience; wells and rest-houses occur at short stages; in many cases the rivers are bridged, and fine avenues of trees extend throughout. The canals are very numerous, but not much used for navigation. They are most common in the east, where the greatest droughts prevail; and when all else around is an utter waste, the contrast presented by the bright patches of green near these "diamonds in the desert," bespeak the benefits they confer on

the land more than tongue can tell. The finest canal is that of the Bári Duáb, which, together with its branches, is 480 miles in length. **Railways** are fast progressing, the completed lines running from Láhor (1.) to Delhi in the south-east; (2.) to Multán and thence into Sindh on the south; and (3.) to Ráwalpindi in the north-west.

[The boats used in the Panjáb are usually heavy, sluggish things, each about 30 feet long, 12 feet broad, sharp at both ends, and able to carry about 12 tons. As all the rivers are very shallow, the boats are made so as to draw very little water; and as the streams are also tortuous, speed is an unknown thing. Generally vessels are allowed to drop down with the current, at the average rate of two miles an hour, or, aided by oars, at the rate of three. Against the current they carry one huge sail if the wind be favourable; if otherwise, they are tracked along the shore like those on the Ganges. When the boats run aground, as they do very often, the boatmen jump out and push them into deeper water. Though the downward journey is slow enough, one day's progress with the stream equals five days' progress against it. Where neither fordable nor bridged, the rivers are usually crossed by ferries; but those who are too poor to pay for the ferry, or too impatient to wait for the boat's departure, either swim, or, seated upon inflated sheep-skins or rafts made of reeds or straw, paddle themselves across.]

273. *Commerce.*—IMPORTS.—**Silk, wool, carpets, madder, assafœtida, fruit,** and **horses**, are the chief imports from Afghánistán and the west; **cotton, woollen,** and **silk cloths, chintzes, metals, glass, cutlery, sugar,** and **spices** from other parts of India. EXPORTS.—These—partly the produce of the country and partly goods in transit—comprise **grain, ghee, hides, wool, shawls, silk** and **cotton fabrics, carpets, cotton, indigo, tobacco, salt,** and **horses.**

[The trade with Afghánistán is mostly carried on by a hardy class of Afgháns called *Povindas.* In summer they are shepherds and in winter traders. Just before the severe weather begins they set out for the plains of India, and visit Multán, Karáchi, Bombay, Dehli, and other large cities.]

274. *Government.*—The Panjáb is governed by a **Lieutenant-Governor.** Each division is placed under the control of a Commissioner.

275. *Revenue.*—This is derived from land, excise, customs, salt, canals, stamps, and sundry minor sources. It amounts to about **three and a half crores of rupees.**

276. *Condition of the People.*—The bulk of the people are very poor, and live in mean mud-built cottages, grouped into villages and small towns. Owing, however, to the increase of cultivation and popular education, the condition of all classes is steadily improving.

[In the remote north-western and western parts the people are still in a very rude state, in fact are but little higher in the scale of civilisation than the Bonairs, Swatís, Wazíris, and other lawless hill tribes who live just beyond the British frontier. These latter, though very troublesome neighbours, are remarkably brave and fine people, and the Wazíris, at least, boast with truth that they "have seen kings come and kings go, but they have never yet seen a king able to take tribute from them."]

277. *Divisions.*—These are very unequal in size, each containing from three to four districts. Their names are **Pesháwar, Ráwalpindi, Amritsar, Jalandhár, Ambála, Dehli, Hissár, Láhor, Multán,** and **Deraját.**

278. *Towns.*—**Láhor,** the capital and seat of the local government, stands near the left bank of the Rávi. It is a large city, containing many palaces, mosques, public buildings, and gardens. In the neighbourhood is the superb tomb of Jahanghir. Near

Láhor is **Mian Mír**, a large military cantonment. **Amritsar** (Sansk. *Amrita saras* = the Fountain of Nectar) is the principal commercial city in the North-west of India, and the seat of the Sikh religion. In the centre of the city there is a tank in which stands a small square temple, covered outside with fretted gold, and gorgeously decorated and inlaid with precious stones within. In this temple, which is dedicated to the Guru Govind Singh, is kept under a silken canopy the book of laws, or "Granth," written by the Guru. **Multán**, near the left bank of the Chenáb, is one of the oldest towns in India. **Derah Gházi Khán** and **Derah Ismail Khán** are places of some commercial note on the Indus. **Attock** (*Atak* = a limit), on the route to Pesháwar and Kábul by way of the Khaibar pass, is situated at the junction of the Kábul river and the Indus. The Indus is crossed here by a bridge of boats, and, as it cannot be used at all times of the year, a tunnel under the bed of the river is in course of construction. **Pesháwar** is a large town, situated in a fertile plain, watered by the Kábul. Anciently it was the capital of Eastern Afghánistán; it is now one of the largest military stations under the Government of India. **Pánípat** is on the great military route between Afghánistán and India. It is famous as the scene of the great battle fought in the plains of *Kurukshetra*, between the Pándavas and the Kauravas, as related in the "Mahábhárat." It is also famous for the two great battles fought in its neighbourhood; the first, A.D. 1525, in which Ibrahím Lodi, the last Pathán king of Dehli, was defeated by Báber, the founder of the Mughal empire; the second, in 1761, in which

Ahmad Abdallah, the first king of Kábul, defeated the Maráthas, and broke their power. **Dehli** (the ancient

View of Simla.

*Indraprastha* = Indra's Dominion) stands on the west bank of the Jamna. It is well known in history as the capital of the Mughal emperors. Anciently it occupied an area of twenty square miles, but great part of it is now covered with ruins. The present city contains several splendid palaces and mosques. The imperial palace is about a mile in circuit; and the Juma Masjid, constructed of white marble and red granite, is the finest edifice of its kind in Upper India. At about nine miles from the city stands the Kutb Minar, a remarkable tower 242 feet in height, erected in memory of a celebrated Muhammadan saint named Kutb-u-dín, in the reign of Shams-u-dín Altamsh, between A.D. 1210 and 1231. Population 160,000. **Simla,** the first sanitarium in India, stands at an elevation varying from 7000 to 9000 feet above the sea. The climate is very salubrious and the surrounding scenery grandly beautiful. The sites of the houses of the residents are scooped out of the hill-sides, as there is scarcely a level spot of ground in the place. There are many more towns in the Panjáb, as **Jalandhár, Ludhiána, Ambála, Siálkot, Karnál, Firozpur,** and others, small in size and otherwise unimportant; nearly all of them are built of mud or of sun-dried bricks, and have dirty crowded streets so narrow that an elephant cannot pass through them.

[The greater part of the Panjáb, as at present constituted, was annexed to the British dominions at the close of the Panjáb war in 1849. At that time the country was almost a wilderness; its highways were unsafe, its resources wasted, and its people in misery. The introduction of good government, police, better education, suppression of national crimes, improved means of communication and canals, have led to increased cultivation and to general prosperity and contentment.]

279. **Protected Sikh States.**—Between the Simla and Hissár Divisions of the Panjáb lie the protected

Sikh states, the most important of which is Patiála. This last is 5060 square miles in extent, and contains one and a half million inhabitants. The soil is fertile, producing large quantities of grain, mostly exported to Láhor and Armitsar. Patiála is governed by a rája, who is perfectly independent, except that in case of war he is bound to furnish the British Government with a certain number of troops. Of late years the condition of this little state has vastly improved. The chief town is **Patiála.** Other protected Sikh states are NABHA and JIND.

280. **Baháwalpur.**—Baháwalpur covers an area of 22,000 square miles, and is one great level tract, of which not more than a sixth part is cultivable. For a distance of ten miles east of the river the ground is covered with luxuriant crops of wheat, rice, tobacco, and indigo, and groves of trees, but immediately beyond the influence of the river's waters the great desert starts up, presenting nothing but an expanse of sandhills from fifty to sixty feet high, thinly covered with stunted shrubs. The people are chiefly Hindus, Baluchis, and Afgháns. The ruling race is that known as the **Dáudpútras** (= Sons of David), so called from Dáud, a chief of note in Sindh, who settled here when forced to flee from his native land. The higher classes speak Persian, the lower orders a dialect of Hindustáni. These last are miserably poor. The revenues of the Nawáb are about twenty lakhs of rupees.

*Towns.*—**Baháwalpur,** the capital, is a collection of mud houses on the left bank of the Gharra.

# KASHMÍR.

281. *Physical Features.*—This tract stretches from the Panjáb on the south and west to Tibet on the north and east, covering an area of 25,000 square miles. Except Kashmír Proper, the whole country is highly mountainous. In the southern portions the hill-sides are clothed with cedar and pine forests and vegetation; but northwards and nearer Tibet there are many tracts destitute of even a trace of the latter. Wherever moisture is, there also is verdure; but where the mountains intercept the rains, the country bears the most desolate aspect.

[The routes throughout this territory are little more than tracks, and are impracticable for wheel-carriages; sheep, goats, mules, and yaks are the only beasts of burden. Often the road lies along the bed of a torrent; often by the edge of a precipice thousands of feet in descent; sometimes the way is blocked up by masses of fallen rock, and sometimes by snow. All the roads have frequent ascents and descents; and journeying, even on the easiest, is always laborious. Places but a few miles distant in a straight line are reached frequently only by travelling many leagues. Streams are crossed by means of a light framework of willow rods, six feet square, resting upon about a dozen inflated sheep or goat skins. These rafts are just able to bear about three or four people. When the streams flow between high and precipitous rocks they are crossed by bridges of planks, or of ropes made of twisted hemp, or grass, or twigs. Though these latter seem dangerous, men and loaded sheep and goats may cross them in perfect safety. Occasionally a bridge is used consisting of a single set of ropes, from which a wooden seat is suspended, and pulled from side to side by a rope worked from the rocks on either side of the river.]

282. *Products.*—As the traveller ascends, he finds not only the vegetation and the climate changing, but also the animals and the people. By degrees the vegetation of the tropics gives place first to that of temperate

climes, and next to that of the arctic regions. Wheat, barley, and buckwheat are the staple productions; fruit-trees bearing apricots, peaches, walnuts, grapes, and apples are common, and form an important addition to the wealth of the people. The elephant, tiger, and other wild beasts of the plain are not found at great elevations; their place is taken by the wolf, the bear, and other similar creatures; and the Indian ox is displaced by the yak.

**[The towns are small and far distant from each other, but villages, consisting for the most part of a few houses surrounded by orchards and patches of cultivation, occur with considerable frequency along the routes. The houses throughout the country are built of stone, or unburnt brick, or clay, with layers of wood. They are two storeys high, and their roofs flat and covered with beaten clay. For windows mica is used, or lattice-work covered with paper. In the higher mountain tracts the houses of the poorer classes are constructed of willow twigs coated with clay or mud.]**

---

## KASHMÍR PROPER.

283. *Physical Features.*—Kashmír Proper is the valley of the Jhílam, here called Bihat, and consists of an extensive plain 4500 square miles in area, surrounded on all sides by lofty mountains, from which numerous offsets on both sides extend far into the valley. Its scenery is of the grandest kind. On all sides the mountains rise above the line of perpetual snow, but those on the north are considerably loftier than the others. The passes by which the valley is entered vary in elevation from 10,000 to 14,000 feet above the sea, and are best on the north-west side.

The flat country or alluvial plain of Kashmír, which forms the middle portion of the valley, is about fifty miles long, not more than ten or twelve miles wide, and is 5300 feet above the sea. It is traversed in its whole length by the river Jhílam, which rises at the east end of the valley, and winds from one side of the plain to the other, at one time washing the base of the northern hills, at another receding to a considerable distance from them. The river flows with a tranquil stream, and being navigable from Islamábád, throughout the whole of the level country, for boats of considerable burden, forms the great highway for the traffic of the province. Wheel-carriages are quite unknown. On the melting of the snow the Jhílam rises higher than the level of the lower grounds, and inundation is prevented by artificial dykes placed along the course of the river. At one time the whole valley formed the bed of a lake; and there are still many lakes and swamps. The soil is very fertile, and produces all sorts of corn in abundance; groves of fruit-trees and gardens are exceedingly common; and flowers of various kinds and of surpassing beauty grow everywhere in profusion, especially roses, which are carefully cultivated for the sake of the attar extracted from them. The mountains are magnificently wooded with forests of cedar and pine almost to their base.

[Among the natural productions not the least valuable is the nut or seed of a water-plant. Many thousand tons of it are annually gathered, and, either in its raw state or roasted, or ground into flour and made into gruel, forms almost the sole subsistence of a large portion of the poor.]

284. *Climate.*—The climate is divided into the

four seasons of **spring, summer, autumn,** and **winter.** Spring and autumn are unhealthy. The periodical rains of India do not extend into the country, but March and April are very rainy, and in July and August thunderstorms occur. May and June are dry and fine. Winter lasts four months, and the ground is then covered with snow.

285. *People, Religion.*—The people are tall and handsomely formed, and of a witty, good-humoured disposition; they speak a language made up of Sanskrit and Persian, and, for the most part, profess the Muhammadan faith.

286. *Manufactures.*—Kashmír has long been celebrated for the excellence of its **shawls** and **firearms,** and amongst the other products of the country are paper, leather for saddlery, lacquered-ware, and attar of roses.

287. *Towns.*—**Kashmír** or **Srínagar,** the capital, is a large town on the banks of the Jhílam, which is crossed by several wooden bridges. **Islamábád,** the town next in importance to Srínagar, stands on the Jhílam at the foot of some steep heights.

[Kashmír became subject to Akbar in A.D. 1589, and he and his successors used it as their summer retreat. They expended immense sums of money in building palaces and forming gardens in it, and, compared with the parched-up plains and burning winds of Dehli, the valley must have appeared to them as a paradise. In 1752 the Afgháns conquered Kashmír and retained it till 1819, when it fell into the power of the Sikhs, who in 1845 ceded it to the British, by whom it was immediately transferred to Ghuláb Singh in independent possession.]

## HILL STATES.

288. The **Hill States**, consisting of BASHAHR, SIRMÚR, MANDI, and several smaller territories, are highly mountainous and picturesque tracts, resembling the hill countries already noticed. Owing to the ruggedness of the country the trade is inconsiderable and the manufactures are few, these latter being, in fact, limited to the smelting of iron and the weaving of coarse woollen cloths for home consumption. The

Mountaineer, Western Himálayas.

roads throughout are little more than mere pathways, practicable only for men and goats or sheep. Bashahr is cut completely in two by the river Satlej; the northern part, called Kunawár, is noted for its grapes.

[The people of these states are subject to the government of their own chiefs, who are all tributaries of the British. They are generally small in stature and very hardy, and in feature resemble the Hindus in the parts towards the south, but towards the north gradually merge into the Tartars of Tibet. They are all ignorant and wonderfully superstitious, but remarkably frank, generous, and truthful. The houses partake much of the character of those of Switzerland; they are substantially built of layers of wood and stone, generally three storeys high. Each storey projects over the one below, and is connected with it by a ladder. In the lower storey the cattle are kept; the middle serves as a granary; and in the upper, which is the largest, the family resides.]

The towns in these states are little more than villages.

---

## CEYLON.

289. Ceylon * is a beautiful island, 25,000 square miles in extent, separated from the Coromandel Coast by the Gulf of Mannaar, and situated south-east of peninsular India, between lat. 5° 56′ and 9° 56′ north, and long. 80° and 82° east. The interior of the island is a tableland, varying in height from 2000 to 3000 feet above the sea, and traversed by parallel chains of hills, with here and there a conical mountain, rising two or three thousand feet higher, and with the most lovely valleys between. All the hills are clothed to the very summits with magnificent forests, and numbers of streams, the largest of which is the Mahávelli Ganga, flow down their sides, giving additional fertility to the rich alluvial soil, which forms a belt about forty miles in breadth all round the island. The northern portion is broken into numerous verdant rocky islets and the peninsula of Jaffna.

---

* Ceylon is called *Singhála* by the natives. In Arabic it is *Serendib*, in Sanskrit *Lanka*, and in Greek *Taprobane*. This last, it is conjectured, is a corruption of *Tivu Rávana* (Rávana's island), from Rávana, the demon sovereign celebrated in the "Rámáyana."

290. *Climate.*—Ceylon is under the complete influence of the monsoons, the north-east monsoon prevailing from November to February, and the south-west from April to September, the intervening months having variable winds or calms. The eastern side is hot and dry, like the countries on the coast of Coromandel; the western side humid like Malabar.

[As in India, more rain falls on the western coast than on the east, and the seasons are contrary; so that on one side of the island, and even on one side of a ridge of hills, the rain may fall in torrents, and the inhabitants may be securing themselves from inundation, while on the other the earth may be parched up, the herbage withered, and the people carefully husbanding the little water left in their tanks and wells. At Colombo the average annual rainfall is 100 inches. Owing to the insular position of Ceylon its temperature is more equable than that of the plains of India. Along the sea-coast the mean annual heat is 80° F., while at Newara Ellia, on a plain 6000 feet above the sea, it is 55°.]

291. *Products.*—The mineral, vegetable, and animal productions are exceedingly various and valuable. Amongst minerals, **granite** is common; **limestone** occurs at Kandi and Jaffna, and **sandstone** along the shores. The precious stones for which Ceylon has long been famous are chiefly **amethysts, topazes, garnets, cat's-eyes,** and **rubies. Pearls** are obtained from the pearl-oyster banks along the north-east and north-west shores. **Nitre, alum,** and **salt** are plentiful. Amongst the vegetable products the chief are **coffee** and **chinchona bark. Cotton** is produced with the greatest facility; every hut has its patch of **sugar-cane** and **tobacco**; **pepper, cardamoms, areca-nut,** and **rice** are produced in abundance throughout the island; **cocoa-nut palms** flourish in the south, **palmyras** in the north, while jack-fruit and other fruit-trees, and many kinds

of the most valuable timber are met with all over the island. **Cinnamon** used to flourish in the south. The animal kingdom is not less rich; earth, air, and water are filled with life. **Elephants** roam about in large herds; alligators haunt most of the rivers, and all the animals of India except the tiger are found.

292. *People, Religion.*—The population is estimated at two and a half millions; and comprises **Singhalese**, who occupy the south and south-west coasts, and are **Buddhists** in religion; **Hindus**, who are in possession of the north and east coasts, and of the peninsula of Jaffna; **Muhammadans**, who are dispersed all over the island; **Veddas**, aboriginal tribes, dwelling in a wild state in the great forests which extend through the centre of the island; and others of mixed European and native descent.

293. *Language.*—**Tamil** is spoken in the north and north-east; **Singhalese** and **Portuguese** are used in the other parts of the island.

294. *Education.*—Education is under a Director, and is making progress, but it is not so advanced as in most of the provinces of India. There is a College at Colombo.

295. *Occupation.*—The people, as in India, are engaged chiefly in agriculture; the manufactures are all very unimportant, except arrack, salt, and oil. The Ceylonese excel in gold and silver work, and in the manufacture of lacquered-ware.

296. *Internal Communication.*—All the considerable towns are placed in communication with each other by fine roads. Those in the maritime parts pass through

groves of cocoa-nut trees along the shore; while those which lead to and from Kandi in the interior pass through a tract the scenery of which is always picturesque and sometimes magnificent. A **railway** runs from Colombo to Kandi, and thence to Nawalapitiya.

297.. *Commerce.*—The commerce consists mainly in exporting **coffee, cocoa-nut oil,** and **cinnamon** to Great Britain, and **betel-nut, timber, salt,** and **coir** to other places, chiefly India; and in importing **manufactured articles** from Britain, and **rice, sugar,** and **tea** from other countries. The chief ports are Colombo, Galle, Trinkomali, and Point Pedro.

298. *Revenue.*—The revenue, which amounts to about **one and a half million pounds sterling** annually, is derived chiefly from land, customs, stamps, cinnamon, salt, excise, and fisheries.

299. *Condition of the People.*—The condition of the mass of the people does not differ from that of the people of Southern India. Their houses are generally of mud, a single storey high, and roofed with tiles, the furniture consisting of a few mats, earthen vessels, and a pestle and mortar; a handkerchief wrapped round the head, and a cloth reaching from the loins to the ankles, form their dress; and rice, curry, ghee, milk and fruit the common articles of diet.

300. *Divisions.*—For administrative purposes, Ceylon is divided into six provinces, each of which is subdivided into districts, each under the control of a Government agent. The government is confided to a **Governor** assisted by two Councils, one legislative and the other executive. For the administration of justice

there are Provincial Courts in the districts and a Supreme Court at Colombo.

301. *Towns.*—**Colombo** is the seat of government and the chief place in the island. Its harbour is capable of receiving small vessels only, larger ones being obliged to anchor upwards of a mile from the shore. A fine breakwater is in course of construction. **Kandi** is situated at the head of a valley 1700 feet above the sea, and surrounded by hills. It was formerly the residence of the King of Kandi, and is now the resort of the Governor. **Newara Ellia**, 6200 feet above the sea, is the chief sanitarium of the island. **Trinkomali** is a small place on the north-east coast, with a harbour capable of containing any fleet in the world in the greatest safety. It is the port of refuge for such ships as are overtaken at sea by the north-east monsoon. **Jaffna** stands on an island at the northern extremity of Ceylon. Much salt is made in the vicinity, and together with the timber of the black palmyra, which here attains perfection, is exported in quantities to Madras, Calcutta, and other parts of India. **Point-de-Galle**, a seaport town, is situated on a low rocky point of land on the south-west coast. It has a spacious harbour, and is a regular calling station for steamers to and from Europe and India, China, and Australia. **Negombo**, **Caltura**, and **Matura**, all upon the coast, and all having some coasting trade, are other places of some importance.

[Beyond the scanty information given in the "Rámáyana" concerning Ceylon, we know little of its history till it was visited by the Portuguese in A.D. 1505. The island was at that time molested by the Arabs, and the Portuguese assisted the native rulers in repelling their attacks. After some years the Portuguese took partial possession of the coast districts, but were expelled in 1656 by the Dutch. The Dutch in turn were expelled by the English in 1796; and in 1815, after a long war with the King of Kandi, the whole island was brought under British rule. A formidable rebellion broke out in 1817 and lasted till 1819, when it was effectually quelled. Since then Ceylon has enjoyed complete peace.]

Buddhist Priests, Ceylon.

# INDEX.*



* Ancient names are printed in italics.

THE END.